NO-BUDGET
DIGITAL FILMMAKING

No-Budget Digital Filmmaking

How to Create Professional-Looking Videos for Little or No Cash

Chuck Gloman

McGraw-Hill

New York   Chicago   San Francisco   Lisbon
London   Madrid   Mexico City   Milan   New Delhi
San Juan   Seoul   Singapore   Sydney   Toronto

The McGraw-Hill Companies

Cataloging-in-Publication Data is on file with the Library of Congress.

Copyright © 2003 by The McGraw-Hill Companies, Inc. All rights reserved. Printed in the United States of America. Except as permitted under the United States Copyright Act of 1976, no part of this publication may be reproduced or distributed in any form or by any means, or stored in a data base or retrieval system, without the prior written permission of the publisher.

1 2 3 4 5 6 7 8 9 0   DOC/DOC   0 9 8 7 6 5 4 3 2

ISBN 0-07-141232-8

The sponsoring editor for this book was Steve Chapman and the production supervisor was Sherri Souffrance. It was set in New Century Schoolbook by MacAllister Publishing Services, LLC.

Printed and bound by RR Donnelley.

 This book is printed on recycled, acid-free paper containing a minimum of 50 percent recycled de-inked fiber.

McGraw-Hill books are available at special quantity discounts to use as premiums and sales promotions, or for use in corporate training programs. For more information, please write to the Director of Special Sales, Professional Publishing, McGraw-Hill, Two Penn Plaza, New York, NY 10121-2298. Or contact your local bookstore.

To my parents whose support throughout the years has been invaluable. Their suggestion to "just keep doing it" made me persevere with my writing and filmmaking. Their most memorable advice—"Just don't do it in the house."

ABOUT THE AUTHOR

Since 1980, **Chuck Gloman** has been an independent producer, videographer, director, and editor with experience in all areas of video from commercial production (800 to date) to corporate training (450 to date). A resident of York, Pennsylvania, he is a regular contributor to *Videography, Mix, Television Broadcast,* and *Government Video*.

CONTENTS

Contents

Contents

PREFACE

It is my hope that this book will be your guide to creating top-notch videos on a very limited budget or "no budget." Besides the obvious cost of tape stock, a video may be made for little, if no other funds. Any video production should stand on its own without having to rely on special effects, an overblown budget, or name actors. What really makes a video good or bad is not the cost involved in the creation process, but rather the quality of the story and how effectively that story is told through the visual medium.

I'm assuming that you have a background in video and aren't looking to learn which end of the camera to look into or which scissors best cut video for editing. These skills are required and need to be learned elsewhere (right after I do). When you are comfortable with the video camera, I will help you learn how to use it more effectively without spending a lot of money to do it.

My background in this ever-changing field began as a child. I grew up watching old-time comedies from the 1920s and 1930s from talents like Snub Pollard, Harry Langdon, Buster Keaton, Larry Semen, Laurel and Hardy, and others who relied on split-second timing, well-thought-out stories (most of the time), and limited production as well as budget techniques.

With very little money to spend to expose the film, these auteurs had to entertain a naive audience. Watching these 8mm and 16mm silent films with my father gave me the desire to do the same thing with my life (not be in old-time comedies): produce films that entertained with the resources I had available (which are frequently none).

I went to the Pennsylvania State University in 1976 and graduated with a master's degree in film in 1982. During that time I made dozens of low-budget films and taught other students the filmmaking process. Upon graduation, a new comer called video arrived on the scene and made film disappear almost overnight.

With my masters thesis film, *The Butler Did It,* tucked under my arm, I literally peddled it to every TV station, distribution outlet, and church group that would watch it. HBO purchased it in 1982 and it remained on the air for nine years.

Finding work in interactive video and training the United States military, I traveled the country learning what video could do on a limited training budget. With a few more years working at an NBC affiliate, I produced over 600 television commercials—each with a budget well under $500. Even the luxury of national commercials only upped my budget slightly. I become know as "the man" if you wanted a commercial produced inexpensively.

Creating videos during the day and teaching video production at a private college at night afforded me the time to pass my low-budget skills (not cheap) to other people who needed to create videos and had no money (students).

What I tell my students is something that I firmly believe myself. You don't need money to create great videos; you need only one thing—desire. If you have the desire to tell a story on the visual moving medium, too much money only clouds the issue and takes your focus away from where it should be. By never having any budget, I was forced to work with only what I had on hand (a kind of sink or swim attitude).

I've seen some great videos, and not one of them cost anything but the video maker's time to do them. There is nothing stopping you from doing the same thing. Unlike the "get rich quick" advertisements you often see, making low-budget videos won't have you driving fancy cars (mine's eight years old and paid for), living in a magnificent villa (I live with Bubbles in a used refrigerator box), or keep you swimming in jewelry (I had my wedding ring soldered on).

But what I promise it will do is allow you to tell any story you want, be proud to show it to anyone, and have the satisfaction of knowing that you did it all without spending a "Hollywood-style" budget.

The following 10 chapters each discuss through anecdotes, experiences, and mistakes what works and what doesn't in a no-budget video. When you have finished reading this book, you can shoot any type of video you choose with less money than you may have thought possible. If I had been given a million dollars to shoot a project, I would have a difficult time deciding what to do with the $999,643 that remained (and they'd have to find me to get it).

Hopefully, this book will not become obsolete as quickly as the camera and editing equipment will be. Whether shooting on consumer analog video, digital, or high definition, the techniques taught here will not change; only the equipment has to change to protect the innocent (or investors).

This book will in layman's terms discuss no-budget video production. At no time will I delve into the complicated, engineering make-up of how the video signal is captured, recorded, placed on the tape, or made a reality. I would rather leave that to the experts and this is beyond the scope of this book (and they know how it's done better than I).

I would rather talk about something that I really know something about —saving money. If you want to learn how to assemble and disassemble a camera in a darkened closet or foxhole, this may not be the book for you. But it you want to make outstanding videos for very little money, this is the book for you.

I'll talk about how to save money in every area of production as well as postproduction. This book is also laid out in no particular order other than the basics in Chapters 1 and 2. If you are only interested in documentaries and never want to see another commercial, skip Chapters 3 through 5 and go directly to Chapter 6 (do not collect $200). I tried to cover all of the various types of videos and how to save money shooting each type.

Once again, if you have lots of money to spend making videos, then share some of that wealth with those less fortunate in this field; maybe they'll name a camera after you.

What Kind of Videos Can I Make?

The Ever-Changing Tape Formats

Since videotape tape was first introduced, the formats have been changing. In the 1950s, videotape was in its infancy. If you wanted to record your favorite half-hour sitcom, you had to hire a room full of equipment and engineers, thread a 90-pound reel of tape that would only last a few minutes, and pray that the pristine black and white signal would actually appear on the magnetic tape.

I won't discuss how the advent of spinning video heads revolutionized tape speed and size over the stationary head format, but let's jump ahead to more recent times and talk about how the video cassette has changed.

While I was in film school in the mid- to late '70s, videotape was a cumbersome reel-to-reel 1/2-inch affair that left a lot to be desired. Two-inch Quad reel-to-reel tape was the industry's broadcast medium, and people were learning how to adjust to shooting 3/4-inch U-Matic out in the field. My first editing project in 1980 involved hitting the Play button on one 3/4-inch deck while simultaneously hitting Record and Play on the other deck. Magically, the image from one tape would transform (although slightly degraded) onto the other deck's TV monitor. Depressing the Pause button would halt the record deck until another shot could be located.

A few years later the *Vertical Helical Scan* or *Video Home System* (VHS), which was begat by JVC, and Betamax recorders appeared on the scene. Let's call this format number one that offered three recording speeds on a standard T-120 video cassette. *Standard play* (SP), *long play* (LP), and *super long play / extended play* (SLP/EP) presented two hours, four hours, and six hours of recording, all in a standard cassette. The other choice from Sony was Betamax, which sported a slightly smaller cassette, a different tape-winding path, and higher resolution images (VHS with 240 horizontal lines, and Betamax with 250 horizontal lines). Betamax also had three recording speeds: Beta I, Beta II, and Beta III. The standard cassette in the Betamax world was an L-750, which enabled almost four and a half hours on the slowest Beta III speed.

Obviously, you don't hear too much about Betamax anymore, even though it had a sharper picture than VHS. VHS won out mainly because people could record six hours of *Baretta* reruns instead of just four and a half on its sharper cousin.

VHS changed little over the years with a slight improvement due to the advent of *high quality* (HQ). The next semiformat change came with making the VHS cassette slightly smaller to fit into a tiny camcorder. This

VHS-C (the "c" meaning compact) opened up the video world with 20 minutes of recording time on the SP mode. This was more than enough time to see Junior spit up his milk for the first time and Wanda Jane singing "Happy Birthday" to Grandma. An adapter enables the consumer to play their VHS-C in a VHS machine and life was once again grand.

Another genius wanted more recording time on a smaller cassette than the VHS-C. He, she, or it came up with the 8mm tape. Slightly larger than an audio cassette that played in your car stereo, this two-hour tape gave the user the same quality and horizontal resolution as a full-size VHS tape without the bulk (just like a Three Musketeers Bar, "fluffy without being stuffy"). Table 1-1 lists analog tape formats with their resolution and other pertinent data.

Being scared, the VHS people realized they needed to do something to make people still want to use their best-selling brand of tape. Almost doubling the horizontal resolution to 400 lines, *Super VHS* (S-VHS) came on the scene with a metal particle tape that would still give the viewer two hours of images with better color separation and sharpness.

Not wanting to be outdone, the 8mm people developed *High-Band 8mm* (Hi8), which accomplished the same thing for the old 8mm that S-VHS did for VHS. Two hours of 400 horizontal-line resolution could be captured on a metal particle tape that was still only 8mm wide. For a long time, this was as far as the consumer could go with videotape.

The professionals, on the other hand, had much more to choose from. Remember that old two-inch Quad reel-to-reel tape? That disappeared overnight with the introduction of one-inch Type C reel-to-reel tape. Half the size of two-inch, one-inch could be edited more easily, offered a higher horizontal resolution of 330 lines, and became the industry standard for editing.

The smaller 3/4-inch U-matic was still a favorite in the field with only 260 horizontal lines because the tape was contained in a cassette shell.

Table 1-1

Analog video tape formats

	VHS	S-VHS	3/4" U	8mm	Hi8mm	1"	Beta-cam	Beta-cam SP
Tape Type	oxide	oxide	Ooxide	metal	metal	oxide	oxide	metal
Luminance	3MHz	5MHz	3.2MHz	3.2MHz	5MHz	4.2MHz	4.1MHz	4.5MHz
Horizontal Resolution	240	400	260	250	400	330	330	470

Even video people enjoyed having tapes in neat plastic containers. The peanut button and chewing gun would not long stick to the reel-to-reel tape and clog the video heads. But as the downsizing trend continued, the people at Sony didn't want to let Beta die.

Betacam quickly became the new industry standard because it sported a 20- or 30-minute tape load on a Betamax-sized cassette. The only visible difference between professional high-quality Betacam and consumer Betamax was the color of the identical-sized cassette and the recording speed. An Betacam cassette ran out of steam in 30 minutes rather than 1 hour and 20 minutes. This happened because Betacam shoved the tape over the heads at five times the speed of Betamax. How's that for letting technology work for you rather than against?

Still not being high enough quality at 330 horizontal lines, Sony wanted a little more money for used Betamax cassettes and followed the same awkward path as its distant third cousin, 8mm. The half-inch tape format now used metal particles and was given the name Betacam *Superior Performance* (SP). Nothing else had changed to the tape except its color, price, and the addition of heavy metal. The lines of horizontal resolution skyrocketed to a whopping 470 lines. Betacam SP at the beginning of the twenty-first century was still the highest-quality analog tape format available.

I did take the liberty of skipping at least eight video formats that rose from the dust and soon disappeared in the ashes. Names like Panasonic's Recam did the same thing with VHS that Sony did with Betamax. Somehow the five-times-normal speed never caught on like Betacam did. MII was another medium, high-quality format that never hung around long enough to make any friends. The other formats like *Extended Definition* (ED) Beta and numerous competitors aren't worth mentioning because the only place you'll ever see them is in museums or at the bottom of land fills. In order for one format to thrive, others must be developed and then disappear. Remember how eight-track tapes enabled listeners to hear tracks over and over? Maybe this is one reason the CD was developed.

Analog video has stagnated and no new developments have surfaced since Betacam SP. The only recent addition to Betacam SP was the introduction of Betacam SX, a slightly improved digital cousin. SX was a trivial metal improvement in horizontal resolution to Betacam SP and a few extra minutes of recording time in a prettier cassette shell.

By far the grandest improvement to the video cassette was the beginning of digital video. The absolute makeup of this superior ally to analog won't be

Table 1-2

Digital tape formats

	Beta-cam SX	D-1	D-2	D-3	D-5	D-6	Digital Beta-cam	Digital S (D-9)	DV-CAM	DVC-PRO (D-7)
Video Data Rate	18 Mbps	172 Mbps	94 Mbps	90 Mbps	220 Mbps	922 Mbps	95 Mbps	50 Mbps	25 Mbps	25 Mbps

discussed here, but some great books out there delve into the nuts and bolts of the format.[1]

First of all, comparing analog to digital isn't quite the same as comparing apples to apples. The terminology is different and without boring you with the great many differences in frequency response, signal-to-noise ratios, sampling rates, and compression rates, suffice it to say that digital is an improvement over analog in every way. Table 1-2 lists the various digital tape formats available and their numbers.

Digital offers a multitude of advantages over analog, such as quality, resolution, copying, freedom of dropouts, and so on. The most pleasant surprise is first noticed when copying one digital tape to another. In digital, the process is called *cloning*, because you are making an exact duplicate of the original in every way, just like a sheep clone. A copy is less sharp and definitely is not identical to the original.

The first digital video cassette to emerge was the Mini-DV. This tiny cassette was slightly larger than a micro-audio cassette and the user could comfortably record 60 minutes on the SP speed.

Since digital video is measured more in data rate than horizontal resolution, let's compare all the digital video formats in video data rate rather than resolution. Most manufacturers still list resolution numbers for digital cameras, but data rate is a better method of comparing each format.

"In digital video the number of bits (four, eight, 10, 12, and so on) determines the resolution of the digital signal. Four bits yields a resolution of one in 16. Eight bits yields a resolution of one in 256. Ten bits yields a resolution of one in 1,024. While influenced by the number of pixels in an image,

---

[1]Silbergleid, Michael and Mark J. Pescatore (eds.). *The Guide to Digital Television, 3rd Edition*. Miller Freeman PSN, Inc., 2001.

the pixel numbers do not define the ultimate resolution but merely the resolution of that part of the equipment."[2] The more information listed in Mbps, the higher quality or more information that picture can contain.

DVCAM is essentially the same as Mini-DV with the exception of recording time. DVCAM runs at a slightly faster speed, taking only 40 minutes to use a mini-DV cassette rather than the usual 60. DVCAM is considered professional, mini-DV is prosumer.

Mini-DV or its immediate neighbor DVCAM has a 25 Mbps video data rate. This is really all that's available at the consumer end of the digital tape market place. I won't be discussing disc or DVD recorders here; that's why this section is called "Tape Formats."

The professionals began with Betacam SX and jumped around from there. Since Betacam, Betacam SP, and SX have been around for some time, Digital Betacam got its feet wet expanding on the ever-present Betacam name. With a video data rate of 95 Mbps, it's a giant leap from DVCAM's 25 Mbps and Betacam SX's measly 18 Mbps.

Not wanting to be left out in the rain, Panasonic's entry in the consumer/prosumer arena is DVC-PRO. Basically the same as Sony's Mini-DV and DVCAM, DVC-PRO's video data rate is also 25 Mbps.

Because digital begins with the letter D, the D series began. The first digital tape format was D-1 (172 Mbps), followed by D-2 (94 Mbps), and then D-3 (90 Mbps). Do you see a pattern developing here? Let's skip D-4 because everybody else did and resume with D-5 (220 Mbps), D-6 (922 Mbps), and D-7, which is actually DVC-PRO. There's always someone who is trying to destroy a great number game, so we finally have D-9 or Digital-S (50 Mbps). The larger the D number doesn't always mean a higher video data rate, but as you can see, quite a lot of digital video formats are out there.

If that weren't enough, someone is trying to change the video standard in the United States from the *National Television Standards Committee* (NTSC), which as been around since 1939, to a new standard called *High Definition* or HDTV as it is commonly called. HDTV currently has three standards to replace NTSC's 525 lines: 24p, 780i, and 1080i (the "p" means progressive scan and the "i" means interlaced).

At press time, a standard has not been decided upon and no camcorders are available for the average person to use and afford (unless you're wealthy). In the near future, we will be shooting on HDTV tape for extremely high quality videos, then we will slowly change to an all disc-based recording medium and tape may disappear as we know it.

---

[2]Silbergleid, Michael and Mark J. Pescatore (eds.). *The Guide to Digital Television, 2nd Edition.* Miller Freeman PSN, Inc., 1999.

There you have it; the semievolution of videotape. The present has brought more changes to video like shooting directly for the Web and recording onto memory cards. The only thing we may be certain of is that as soon as you buy a new piece of equipment, it is already obsolete.

Equipment: One Chip Versus Three

With hundreds of cameras out there to choose from, how do you know which one is best? The answer to that question may be simpler than you think: whichever one works best for you. Without being wishy-washy, you want a camera that serves your needs, is comfortable, and is easy to operate.

I won't describe all the models and prices of camcorders because they change too quickly. As long as you know the types and functions of the equipment out there, you will live a longer, happier life.

The choices of cameras fall into two categories: one chip (consumer) and three chip (prosumer or professional). Which camera you decide to use really depends on your budget, and since this is a book about low budgets, we'll focus more on the consumer one-chip models. Whether you use a JVC, Sony, Panasonic, Hitachi, or any of 30 other models, one-chip cameras basically capture the image on tape or disc using a simple, several hundred-thousand-pixel *charge-coupled device* (CCD).

After having shot over 2000 "videos," I've used almost every model and type of camera out there. You have a choice of recording your image on three mediums: tape, disc, or memory card (chip). The tape formats available alone are enough to send your head spinning. Most old-timers begin with VHS. This hefty camcorder enables two hours of recording time on a standard VHS T-120. Obviously, longer cassettes or slower recording speeds offer epic proportion programs, but most want to use the highest-quality SP speed before breaking the record tab. This type of setup has its problems.

Weddings

Let's use wedding as an example and see which works best to tape your sister's, niece's, daughter's, or ex-wife's wedding. These cameras rarely function well in low light, so that romantic candle-lit toast was basically just that: a grainy toast. The camera is bulky and becomes a burden by the time the reception starts. The battery supply needs to be changed several times during the tape and before the advent of VHS HQ (about 1996) the images were muddy, grainy, and slightly less than 240 horizontal lines.

VHS camcorders also have a few advantages. They are cheap, most costing less than $300. At the end of the event, the VHS tape could be handed over to the happy couple and they could watch it in any standard machine. Wedding guests will always think you work for some news station because of the monstrous size of your camera.

VHS-C is obviously a smaller form of VHS in an effort to make the camcorders into more comfortable palmcorders. The next tiny leap in quality is 8mm. Roughly the size of an audio cassette, 8mm shrunk to the size of the camera; with the battery installed, it weighs in at slightly over a pound. Able to compete with the heavy weight VHS, 8mm offers the same 120 minutes of recording size with the substantial horizontal-line image improvement of 10 lines. At 250 lines of horizontal resolution, you could now see the fine crumbs on the white tablecloth.

The sizeable video-quality increase kept the camera the same size but almost doubled the horizontal resolution to 400 lines. The 8mm tape also changed its name to *High Definition 8mm* (Hi8) and accomplished that by using a metal particle formulation tape. Wedding videographers were thrilled that they now had extremely high quality accompanying a smaller camcorder. If the couple wished, the videographer could follow them on their honeymoon without many others noticing.

A more recent trend in one CCD chip technology is the digital camera. Instead of recording the image in an analog format, ones and zeros do the same thing and actually take up less video real estate in the process. Digital offers higher resolution (450 horizontal lines) and no dropouts (those nasty white flecks that appear on the screen where video isn't) at the price of a shorter 60-minute tape speed in the highest-quality mode.

The latest camcorders record the images on video CD, mini DVD, or onto a memory card, but the camera has not changed in size. Only the price has changed to protect the innocent.

The next tremendous leap in quality occurs when you transform your shooting from a one-CCD to a three-CCD camera. Even if you are bad in math, three CCD chips, each containing several hundred thousand pixels, will look far superior to a camera that just has one. With three chips doing the same job as one, the chromance and luminance (the color and the light level) are sharper and more defined. If you had to draw a picture with one crayon or three different crayons, which one do you think would look best? See, now you can tell you friends that this is an interactive book!

To even the seasoned amateur, the three-CCD camera's image is far more professional looking. The cameras are slightly larger than their one-chip cousins (you have to make room for the extra chips, like a family-sized bag of potato chips) and most three-chip cameras are digital.

The biggest leap is in the size of the chip. Prosumer cameras costing under $3,000 incorporate either 1/4-inch or 1/3-inch CCD chips. With your improving math skills, you may notice the larger the chip (1/3 of an inch being larger than 1/4 inches), the sharper the image and usually the greater number of pixels. If you decide to purchase, rent, or borrow a camera with 1/2-inch or 2/3-inch CCD chips, you have reached the big leagues in the professional world. However, now you may be shelling out over $10,000 for your device. It will take much better pictures, but how do you pass that cost along to Heckle and Jeckle getting married? If you rent yourself out with the camera by the day or the event, you need to recoup your cost for the camera and the budget immediately goes higher. My advice is to not use the expensive models and stick with one- or three-chip cameras. Figure 1-1 highlights a popular three-chip camera compared to its younger, cheaper, one-chip cousin.

Weddings still look great with the new digital, one-chip cameras. Your cost, other than the camera, is only the videocassette. If you are shooting a two-hour wedding, have at least four times that amount of tape. If you don't use it, another client will also be able to.

The best advice I can give you is just try as many different models as possible and see which one you like the best. Try to borrow someone else's

Figure 1-1
Three-chip camera
versus one-chip

camera and shoot some video with it. Is it comfortable? Are the features easy to access?

Even when looking at cameras in stores, if it is uncomfortable to hold, it may end up sitting on a shelf never getting used. Remember only you can be the judge (and help prevent forest fires).

What's Out There and How Do I Get Involved?

How do you tap into all the video work that's waiting for you? You may even know of productions happening in your town, but how do you get involved? That is easily answered: Just show up. I'm not telling you to crash every video production and begin asking a million questions, but if you make yourself available and let people know your abilities, the job becomes much easier.

If you ever visit the Mecca of East Coast production, New York City, or on the West Coast, Los Angeles, you will see a production happening on nearly every street corner. It's usually acceptable to watch these productions, but some don't allow you to walk up to Brad Pitt and tell him a better way to act in the scene.

Each type of video production (except one) is listed or discussed in this book. Like magazines, videos are being made on every subject on earth. Visit any video rental store and look in the specialty section. More on these types of videos will be discussed in Chapter 5, "How-to Videos."

The only sect of video that won't be covered or uncovered in detail is the porno video. Pornography, especially on video, is a multibillion dollar business. With over a hundred million camcorders in the United States alone, making videos is downright easy. As the title of this book implies, *No-Budget Video Production* really hits home with video porno. I've read that the production values aren't too high, the sets inexpensive, and the acting amateurish. Basically, the budgets are pretty low.

I believe this market is best left alone by the budding video artist. Once you associate yourself with shooting, editing, or duplicating pornography, you are pigeonholed in that genre. The money is fabulous, the hours are short, but it will stay with you the rest of your life. I'm strictly talking about the "legal" pornography involving only adults.

I noticed an ad in the classifieds for a "Videographer with Own Equipment." I called the Florida phone number and left a message. Two days later

I got a response and was asked if I would answer a few questions. Believing they wanted to know what type of equipment I had (camera equipment), my daily rate, and so on, I was ready with a listing of model numbers. I was way off base. They first asked me what I liked most about pornographic videos. I truthfully told them "nothing." They asked me squat about my skills as a videographer, but every question involved pornography. It seems they wanted to hire someone with *any* type of video equipment (consumer or professional) to shoot in a given hotel room, in a given city, for $200 a day plus expenses. Magnum P.I. charges more than that and he drives a Ferrari. I told them I wasn't interested.

A colleague owns a video production business and was asked if he wished to edit pornographic video. He was offered $200 an hour to edit someone else's footage. He immediately said no even though the money was good and business was slow, all because he had a conscience.

A close friend of mine had a 35mm film production business. He was brilliant in that he shot, directed, and edited everything himself. He even went as far as processing his own film stock and marketing all his features himself to local theatres. His genre was X-rated movies (long before video was released). Here was this extremely talented filmmaker who peddled porno at $6,000 per feature film. His budget was low, his production values excellent, and his 35mm motion picture equipment pristine. He died without ever doing any other type of work; he was far too talented to waste his time, but he was typecast because that's the type of work he did.

I'm not telling what type of video work not to get into, but if you decide to make the big bucks and shoot or edit this type of video, that's all you will ever do. The choice is yours and you will have to live with it.

If you don't have a lot of experience in making videos and wish to become more knowledgeable, there are only three ways: reading about it, going to class to learn about it, and doing it. The best way is probably a combination of all three, but let's break each one down and see what's involved.

Reading Makes the Heart Grow Fonder (Book Learning)

Subscribe to every video magazine you can get your hands on. The magazine racks have a few monthly issues on the shelf, but check out this book's Appendix for some of the professional trade journals. Bookstores also have tons of books on the art of video. Read all you can because you can never be too knowledgeable (then you would be a know-it-all).

Read how others in the business are doing videos: What can you learn from them, do their mistakes help you create better programs, and where are they getting work? Ads in magazines, job postings on the Internet, or even the local paper are some places to find video work.

Two of the best places to find work matching your skills on the Internet are www.mandy.com and www.cinematography.com. Both of these listings offer hundreds of jobs throughout the world in video production. Most of these listings require experience, but some need low-cost interns who want to learn. Hundreds of sites exist on the Web; choose your favorite browser and take it from there. Look at Figure 1-2 for an example of Mandy's web site.

Get Back to School (Class Learning)

A great way to learn, sometimes without having your own equipment, is to listen to teachers or experts. I teach video production in a private college. As a producer by day, I bring my real-life experiences into the classroom in the evenings (I'm afraid of light). Students get hands-on skills with the equipment and also learn the "why" behind each concept. The basics need to be learned and mastered before moving into the professional marketplace.

The downside to classroom learning is that it's time consuming and sometimes expensive. I often attend classes or seminars to hone my skills

Figure 1-2
www.mandy.com

in new techniques or equipment. This process takes book knowledge one step further by making it interactive, which is a higher form of learning.

Students must work on their own as well as on student projects. Again, this hands-on approach enables them to learn from their mistakes without being punished by the client. I often have my students help on my shoots (in paying roles) and I find they work better than some professionals because they are still excited about the business.

This is another place to save a few dollars on your productions. Check out local colleges, community colleges, and video clubs for raw talent. Most of these individuals have enough drive, enthusiasm, and often the skill at a much lower price than "professionals." I might not hire a 14-day video veteran to direct my next feature (although it would save a few dollars), but most projects have numerous roles that video students fit into nicely. My students enjoy just being on the set to soak up the atmosphere. Remember, keep your budget low by hiring video makers in training. Using students, as shown in Figure 1-3, will make your crew look bigger than it really is.

Interns, as the name implies, are professionals in training. Internships may be paying or nonpaying jobs, but both get you where video is being produced. In the beginning, you might be more concerned about what you learn than how much you will make. Most of my students always want to know

Figure 1-3
Students working on location

how much money they can make when they "get out." I always tell them "as much as you want." I have no idea what they want to do with the rest of their lives (I could always ask them, but that would be cheating). The more video jobs you take, the more money you will make; it's the same as in business.

Focus is the word of the day. Focus on what you want to do and do the best job you possibly can at it. Persistence will pay off, but first you should pay your dues.

Just Do It

The cheapest and most effective way of learning is by doing. They say those that can, do, and those that can't, teach. If you want to learn how to do a new task, the best way is to do it. How do you know if you can hold a shot steady, edit a tightly knit commercial, write a moving script, or direct the next blockbuster unless you try by doing it?

Volunteer to work on projects until you gain the experience to work solo. Very few video makers will throw you off the set if you have a desire to learn, don't ask a billion questions (don't get in the way), or can actually help the process. Most of the feature films I worked on were learning projects. By watching the experts at their craft, I learned what works and sometimes what doesn't.

Television stations always need new blood behind the scenes. At least you will learn the fast way to do things on live television and focus on what can be done for little money. Don't be afraid to assert yourself and ask if a local station has any openings.

Make videos on your own with friends and family. All you need is a camera and tape and you can begin to complete your first no-budget production. I'm sure you're full of ideas (I'm full of something else); make those ideas a reality. If you want to shoot videos for a living, the more you shoot, the better you'll become. Offer to videotape every event your family or friends are involved in; this teaches documentary filmmaking.

If possible, tell a story by cutting in the camera; this teaches editing. Explain to the actor(s) exactly what you want them to do. This teaches directing.

When I first wanted to enter film school at Penn State, I was told I had to shoot a film (no video yet) so the department could determine if I was worthy to enroll. In the summer of 1978 I wrote, shot, directed, and edited a 45-minute Super 8 film called *Forever Was Yesterday*. Friends and family helped throughout the process and at the end of the summer I presented the humorous epic to the college.

They touched the can of film briefly and said, "We don't need perspective students to shoot a film for enrollment anymore." I had just spent the entire summer and a lot of money shooting this masterpiece and now they wouldn't even look at it. But the experience I gained in the process was worth every minute and dollar I spent (okay, I borrowed the money from a friend).

How Cheap Is Cheap?

Try to look at shooting no-budget videos as a game or challenge. How can you produce the best-looking video for the least amount of money without skimping? After all, no one has any money to spend on a video, and once you tell your client the cost, they suddenly lose interest.

Unfortunately, once you set a price for yourself, everybody tries to get a special deal. Let's look at an example of pricing to see how cheap the word cheap can be.

A local church performs a concert to the congregation and general public once a year. On their twentieth year, they called me and asked if I could put together a compilation of the "best of 20 years." This is obviously an editing job: looking through 20 years of someone else's footage and deciding what stays and what goes in the anniversary tape.

The first thing every client who calls asks is, "How much does it cost to do a video?" I immediately reply, "How long is a piece of string?" There is no way to tell someone how much a video costs until you learn a little more about the project.

When I was a car salesman in my torrid youth, some customers wanting a car were only concerned about the price. They would walk up to me in my nice plaid suit and say, "How much?" Knowing they weren't asking for my body, I assumed they were speaking about the automobile they were looking at. No matter what I tried to tell them about the car the only thing that was pressing on their mind was "How much?" These types of people only want to hear one thing and that's the only way to satisfy them.

At the dealership, only the manager was allowed to set the price. Knowing the persistent man wouldn't believe that, I just gave him a price. That's right; I made up a high price. I figured if he didn't want to play by the rules I would just give him the first high number that popped into my mind.

Along the same lines, I could quote the church lady the same formula number: $1,000 a finished minute. Of course, she would thank me and promptly call someone else. If people want to know a price, that's what the price is, but every video ever made is different.

As soon as they ask, "How much?" I immediately ask several magic questions. The answers to these questions allow me to formulate a reasonable price. The questions are (in no particular order): what would be involved in the project, what is your time frame, what is the final running time, and, most importantly, what is the budget? When you ask all these questions, you are much better prepared to tell them how much and be closer to the mark. With all these questions answered, you are better prepared to give them a price.

You need answers to at least some of these questions to deduce a fair, cheap price. If the perspective client doesn't know the answers to *any* of those questions, they probably aren't ready to spend any money on a video; they may just be shopping around for prices. These people are called "tire kickers" in the video and car business.

What's Involved?

Knowing what you are committing to is a great starter topic. Every client should be asked about this. As soon as you ask it, shut up and let them talk; do not lead the witness or badger them. They probably haven't really thought any of this through, so this may be the first time someone has asked them an intelligent question about the project.

Knowing what's involved, or more clearly asked, "What would you like me to do on this project?" helps you determine how much time is needed in preproduction through postproduction. Do you have to come up with the creative (the script)? Do you have to please just you or a group of people (how do you please everybody)? Will you be the producer, director, and editor, or will someone else be making decisions? These subquestions will determine your responsibility and who you may have to answer to.

The church lady, when asked this question, wanted me to look at all the tapes and determine what was "great stuff" and what was "trash." An outsider like me would have a less biased view if some footage was good or bad because none of my precious siblings were involved.

Ten hours of VHS tapes would obviously take at least 10 hours to work with, probably more like 12. I would view the material, marking the good footage and compile a paper cut. This task would be charged by the hour. Come up with a fair hourly fee and multiply that by the number of hours needed:

Hourly fee ($) × Number of hours = Initial cost

If this amount is too high for the client, you've saved yourself a lot of time and effort. The church lady approved step 1.

You Want It When?

This might be the most important question to ask. If the client has an unrealistic schedule and wants it completed immediately, you better know that walking in. If you have the time and are willing to work quickly, you may still accept the job. Just keep in mind how this got to be a rush project in the first place. Probably it sat on a back burner too long, everyone else they called wanted too much money, or this is just someone's whim and they want to see if it can be done quickly and cheaply.

Accepting projects like this always scares me, even if I am in a slow period. A rushed project means cut corners, the client not really knowing what they want or even like, and extremely long hours. Few outside the business realize that a 10-minute video takes longer than 10 minutes to create. Luckily, the church lady knew that watching all the footage was going to take time and the project would be needed three months down the road. I was exceedingly fortunate to have that long to complete the project. They wanted it done right and could afford to take their time.

How Long Is It?

The final running time of a video is always a gray area. Some say they want a 10-minute video when a 5-minute opus is better suited to their needs. Whatever time the client tells you, either cut that in half or multiply it by three.

If you are compiling a video from other footage that people shot, expect the running time to be a lot longer than they quoted you. Once they watch the footage, they will want to keep almost everything and cut very little. This will add greatly to the running time. Sometimes they have a specific time in mind because that's all they want. The church video could only be 30 minutes because that's all the time allotted for the captive audience.

If you are shooting a video from scratch, it will be a shorter project than the client expects. If someone requires you to shoot an hour video on something, a 20-minute project will be cheaper and more interesting in the end. This comes with experience, but I've rarely been wrong with my two formulas:

Compilation Video

Client's proposed total running time (minutes) $\times$ *3 = Actual finished running time*

Story Video

Client's proposed total running time (minutes) $\div$ *5 = Actual finished running time*

What's the Budget?

Better yet, you should ask, "How much are you willing to spend to get this done?" This will be the most difficult question for them to answer because most have no idea what they want to spend. If you instantaneously told them "how much" when they first asked, they may not know if it was a good price or not.

This is really a sneaky question to ask in my opinion. You are actually telling them you want to know how much money they have so your price will fit without you having to guess. Just tell them to put the money directly in your pocket. If I only ask one question in video, this will always be the question.

If someone calls out of the blue, you have no clue what price they may have in mind. If you ask them what their budget is (or more crudely "How much money do you have to spend?"), they will tell you the magic answer. If a new client calls and their budget is $5,000 for a five-minute video, you can charge your normal rates because that's $1,000 a finished minute. Of course, this rarely happens. If the same person says $1,000, can you accomplish that with the same equipment as well as your time and skills?

Let's look at another possible scenario. Someone calls and wants to know how much to produce a 10-minute video. If you blurt out a price (only after *all* the other questions have been answered), you are stuck to that price. If you tell them $5,000, they know how much the project will cost them. If you ask them their budget or price range and they say $1,000, you no longer have to tell your price because it is five times what they want to spend. Politely tell them that you cannot do the video for that little (unless you are desperate for the work).

On the other side, if you ask them their budget and they say $10,000, aren't you glad you didn't say $5,000 right away? That doesn't mean you have to spend all their money, but at least you have the luxury of not pinching pennies. Always leave the ball in their court and feel them out so to speak. If the price is too low, maybe you can recommend someone who could do the project for the cheapskate, I mean, client.

Asking the budget really saves you from underbidding a project. I have no trouble overbidding a video, but I never want to say I'll do something for $1 when I could have charged $5 for the same thing. Even though I've been quoting prices for well over 20 years, I still always ask these questions so I know what I've just sold my soul to do (you can do that with a church project).

Now that you have asked all the pertinent questions, you can reasonably give the church lady a price. If possible, don't quote a price over the phone immediately. Take your time and really do the math to see how much you should charge. The church lady kept screaming for a ballpark price. At a ballpark, a hot dog costs $5 and isn't worth that price. Instead of giving her a ballpark, give her a range.

Using all the rules of the game you've learned and will learn, give them a several thousand dollar range. You have no idea what you are walking into. All the video could look like garbage and it may take weeks to clean everything up. The video may be wonderful and you finish in a couple of days. Tell them the truth. Until you know more about what's involved, the best you can give them is a range, which in this case would be $1,000 to $5,000.

The project just involves editing, but you don't know how much. The church will never pay $1,000 a finished minute unless they have very deep pockets, and realistically it will cost at least $1,000 for your time and editing. This is the hardest part of doing videos. You are a professional, and professionals are paid when they work because they are using their specialized skills. Besides, you have to pay for that Yellow Pages ad somehow.

Speaking of church projects, as soon as people know you are a video person, you immediately become the authority on everything involving videotape. My church (where I am a member) knew this and asked me to produce, from scratch, a 30-minute documentary for the year 2000. Several members would talk on camera and I needed to produce, script, shoot, light, direct, record sound, edit, and duplicate the production before January

2000. I was alerted to this fact on Thanksgiving 1999. I've done projects in less time, but it was my duty to make this happen.

I asked the standard questions and each one was answered in due time. When all the information had been collected, I knew exactly how many hours I needed to complete the job. I also knew exactly how much it was going to cost the church. Camera rentals, editing room rentals, and tape stock had to be acquired and it took dozens of hours of my time to pull everything together.

Cutting every corner and making deals, I felt I had a cheap budget to present. Their idea of cheap and mine were very far off. Since I was a member, my services would be free. Okay, that wasn't the price I was looking for, but I was doing this for God. God didn't ask me directly about the project, but He did give me the ability to create videos. I wasn't going to wrestle that issue. All the rental equipment cost me money to acquire, so that would be paid by the church. If I hadn't mentioned that, I would also incur those expenses. Always be truthful, even in church.

If you don't have a piece of equipment and need it to get the job done, tell the client that it is an expense for you. Most people rent this equipment and mark it up because of the time, travel, and the expense to get it (I couldn't do that for the church). The equipment I owned, the lights, and sound equipment were also donated. This is what we call a freebie. I was also able to find great deals on the rented equipment. Because a church was involved, I was given lower prices and I passed that directly on to the church.

My reason for telling this story is sometimes you need to do really cheap or no-budget productions. You can't make a habit of this or you'll starve, but it is good public relations for a couple of reasons:

- You get practice in honing your skills in production/postproduction.
- Doing a charitable project helps you as well as the community, and lets people see what kind of work you can do.

Everyone in the church knew I did videos, but they assumed (you know what happens when you assume) I just used a camcorder and shot home movies. When the final video was projected in front of the congregation, all were amazed at its professional quality. They never expected moving camera angles, modeled lighting, dissolves, character-generated text, music, and voiceover narration. I used professional equipment and worked to the best of my ability.

Afterwards, I was greeted by no less than 50 people who were amazed at the quality of the production. Not only does this make you feel good as a professional, but I got numerous leads for other videos that people needed at their jobs. "I had no idea you did this! We've got a 40-minute project that we need you to make."

Because of a freebie, I got four additional lucrative projects that kept me off the streets for quite some time. If I had said no to the church project, I would have lost a lot of work. Think about the possible referrals when you do a project. I never believed the church video would advance my career, but good things come in small, sometimes unexpected, packages.

Event Videos: Avoiding Pigeonholing

To some people, the words "event videos" make them cringe and believe the individuals who pursue this line of work are slightly less than human. After all, how could someone possibly make a living videotaping other people's events?

First, let's define the term event video. An event is simply something that happens, be it an occurrence or an event. No one wants to call it an occurrence video because it sounds like some Wild West showdown. I won't bother defining the term video because hopefully you already know what that means.

Typical events are weddings, bar mitzvahs, plays, recitals, parties, and almost anything else that someone may want taped. Events like funerals and autopsies may be recorded, but that's not what I call an event that someone would want to see. We'll discuss documentation videos later, and such events fall into that category.

The most popular event videos shot today are weddings. In a wedding, everyone is happy (unless they had been bickering the night before), everyone is well dressed (nudist weddings will not be covered here), and generally it's an atmosphere of euphoria (of course, the bride and groom will have no clue what they did the next morning). Figure 1-4 is an actual video image from a wedding shot on a three-chip camera.

For a moment, envision what photographers must pay up front before the wedding. They have to deal with the cost of film stock, processing, and printing the images. Even digital still photography has more up-front costs than video, all without sound.

Now that you have chosen a camera to shoot the wedding, in order to keep the budget down, try to edit in the camera. By that I mean don't try to record the entire two-hour wedding service unless the couple desires that.

Figure 1-4
Wedding video

Using time compression, just shoot the highlights of the day. If you end up shooting too much or you see Aunt Ethel drunk and flirting with someone's great aunt, editing is the only way out of the mess.

Transferring electronic images from one tape to another or editing on a nonlinear system still incurs costs. With careful planning, you may edit the entire wedding in the camera without ever having the need to edit your work. With practice, you'll see how easy this is to accomplish.

Some couples even want two cameras to cover their fateful day. This means adding twice the headaches and twice the costs (camera rental and editing). In order to keep costs down for both parties, try to keep with one camera coverage.

The only other cost in this massive budget besides tape stock and the pressing of your best suit is the camcorder. We all know people who have cameras and with the cheapest wedding videography costing at least $200, a camera could be purchased after two events. Borrowing is also an option; just make sure you return it for someone named Bubba is involved.

Bar mitzvahs are much like weddings except that the groom isn't married, he's probably Jewish, and he's a little young for most women. The process of shooting this event is the same as the wedding: edit within the camera. Whether shooting a wedding or bar mitzvah, attend one before you begin shooting so you are comfortable with the arrangement of what happens when. They always have wedding rehearsals before the big day. Why not attend it and practice your skills?

Plays and recitals I lump into the same category (what some call boring). Whether you are taping your kids, relatives, friends, or strangers, filming a staged event is difficult. Unless you've heard little Wanda Jane rehearsing her part for three weeks, you probably have no clue when someone is going to speak, leap, fall, or run. Attending rehearsals or dress rehearsals allows you to see and make mental notes of what is happening. Some videographers shoot the entire performance in a wide shot so everything will be captured. That, again, is the amateur's way of doing things. Part of telling a story on video is by using close-ups, zooms, and other angles to make the uninteresting less so. Editing in the camera lets you zoom to a dancer's feet, pan to another character, or zoom in for that touching moment in close-up.

It takes perfect intuition to know what every performer is going to do and when, so try to stay alert and follow the action. You will make mistakes, so don't immediately snap the camera back to your previous shot when you follow a bad lead. No one may know you made a mistake if you simply and slowly pan back to or zoom out to a more pleasing shot. You still should get at least one close-up of every stage person because Mom and Pop want to see Junior or Junette doing what they do best, making their parents proud.

If charging for the event, begin with a long shot to establish the scenario, then slowly zoom in, and pan left to right or right to left to get close-ups of every person involved. Every parent will see their precious little ones and you will have covered the whole event. In long lines of dancers, pan to the left and then right slowly, so all people are shown. If you miss someone, that's the only time that person's parents will complain. Remember, keep all your moves slow and deliberate. Don't do anything too quickly (just like eating ice cream too fast).

If you are being paid to be an event videographer (if you do it for nothing, then it really is a no-budget production), your clients are expecting a professional production. Try taping a few events for no charge until you feel comfortable with your skills as a videographer. A practice video of a wedding is a great gift, it allows you to gain valuable experience, and they will love it because it was semiprofessionally done.

By shooting events, you will quickly learn that this isn't the field for everyone. The hours may be very long, you will have to put up with people who believe they are always right (because they're paying you), and you must be dependable. If you get a higher-paying gig right before your intended event, your old clients are expecting you to be there. The same goes for bad weather, sickness, dark of night, and incompetent transportation. The most important thing to remember about event videography besides punctuality, neatness, and smiles is your equipment. The paying public is expecting everything you have to be functioning properly on the big day.

If the camera craps out, eats a tape, or bursts into flames, it's your responsibility to get it working again or find a replacement—quickly. A good rule of thumb is to expect the unexpected and bring along a backup camera.

It doesn't have to match the other camera in any way (except that it works). This backup camera is just to be used in a pinch; you are not doing the event with two cameras (more footage, more editing, and more money). Even if you consider yourself an amateur, a backup camera is a professional thing to do.

A friend of mine shot a wedding with a $50,000 Betacam SP camera with 30-minute tape loads. He spent the entire day following them around and recording everything. At the end of his 8-hour commitment, he had 16 tapes that would be edited down to a 2-hour project. When he played back his footage, he had 16 tapes of nothing—blank footage. Can you imagine what he felt when he first discovered this problem? Or more importantly, what he first said? Because the harm was already done, he suggested making their still photo album come alive as a video. Since he couldn't reshoot the wedding, he scanned all the still photos and edited it to music. It's not the same as a wedding video, but the small costs he incurred helped turn a tragedy into just a lesson learned.

A professional or good amateur uses video because it has the capability to be played back immediately. Film photographers can't do that, but videographers can. Even if you only playback a few seconds of material, just make sure you are recording something on the tape or disc.

This brings me to another point. Because camcorders are silent, you hardly know they're recording when you press the little red button. I usually turn off the red tally light when shooting events because it's better for people not to know when the camera is running. I've had too many "deer in the headlights" when the red light was on. If you can't turn yours off, put a piece of black electrical tape over it.

With the red light covered, you are the only person who knows when the camera is recording. Have you ever hit the record button to begin your epic but not pressed it hard enough to activate the mechanism? When you were satisfied, you press the button again to stop recording, only to actually begin recording. I have shot quite a lot of videotape of my feet, people's stomachs, and the camera being carried around from place to place.

When you activate the Record button, make sure you are actually recording. Look for the "Rec" indicator, footage counter, or some other clue that the tape is moving. After shooting for over 20 years (boy, am I tired), I still check, check again, and triple check to make sure that I am recording when I want to record. All the footage of my feet I occasionally get when I fowl up is edited out, increasing my budget.

Now you should know all there is to know about event videography. I know people who make a great living doing it (I still do it occasionally) and don't mind the repetitiveness. But every event is different and if you look at it that way, you may have found a new money-making endeavor.

Legal Videos

A great opportunity for wealth with a limited spending budget may be obtained in the vicinity of legal videos. If you really want to get picky, all videos created are actually legal in the fact that they exist and are real. Now that my attempt of humor has totally flown over your head, what exactly is a legal video?

A legal video fits into four categories: deposition, day in the life, mock trial, and documentary. As the name implies, legal videos are usually done for attorneys because they are the only people on earth who find these programs remotely interesting. Actually, you can make these docudramas somewhat entertaining, but the purpose of shooting these is to win money or a settlement for the party involved. But don't let that stop you from being creative. If you don't have fun with what you shoot, it becomes stale.

Most of these programs do not require huge amounts of skill on the part of the videographer. You cannot hire a chimp to shoot these programs, but dependability, preciseness, professionalness, and following the rules outweigh the skill factor slightly. Before the legal professionals out there get mad at me and say I don't know what I'm talking about, let's look at what's involved in shooting some of these videos.

The Deposition

The most mundane, but sometimes the most important, legal video is the deposition. A lawyer is usually deposing or questioning a perspective alleged witness (see, I really know how to use legal terms) to recall some particular events. Before the advent of our magnetic friend called videotape, a stenographer typed every word uttered by the witness on a nifty little machine. Depositions are done because lawyers want to know what a person is going to say on the witness stand before he or she gets up and embarrasses everybody by saying something new. You've all seen Laverne and Shirley on the witness stand; see how important depositions are? The lawyer doing the questioning in the deposition is going to ask everything the other attorney may quiz the badgered witness on. If you know beforehand what Junior is

Figure 1-5
Deposition video

going to say on the witness stand, you won't be in for any surprises (like Laverne). Look at what little is involved in the setup of a deposition (see Figure 1-5).

If cousin Vinny calls his deposed witness on the stand and she utters something new and revealing to the same question that was asked in the deposition, Vinny may read from the deposition transcript and quote exactly what she had said earlier. If you need more information on these legal matters, read one of John Grisham's books or watch *My Cousin Vinny* a sixth time.

Now that video exists, the stenographer still records every word, but video can record the facial expressions, body language, shifty eyes, and quivering voice. The less done from the videographer, the more successful the deposition. You need to let video do what it does best: record reality.

Depositions may last a short period of time or several hours. The key is that the camera is stationed on a tripod and is not panned, tilted, or zoomed during the shot, and the significant things are recorded. Let's do a hypothetical deposition for a top-notch attorney.

When you are hired to shoot a deposition, the most important thing is to be punctual (we'll talk about the money later). If the deposition begins at 9:00 A.M., be ready to start taping well before that time. Lawyers are charging hundreds of dollars an hour for their time, the person being deposed has left his or her job and is not getting paid, and the opposition lawyers will also be present. Depending on which area of the country you live in, there may be over $1,000 an hour just in legal salaries riding on this video.

The person being taped is sitting, so the tripod should be at his or her eye level. The type of camera isn't important and, depending on the budget, discuss with your attorney if a one-chip or three-chip camera is warranted. When I shoot depositions, I usually use a one-chip camera to keep costs down since the lawyers are more concerned with the words spoken and body movements rather than pristine video sharpness.

"The longer recording time on a cassette, the better deposition you will make," Yoda believes. It is very distracting to have to change a VHS-C every 20 minutes. A two-hour VHS, 8mm, or DVCAM is the better option in this case.

A microphone on the camera will record the sound of the lawyer as well as the witness, but it will also overmodulate on every cough, sneeze, or slamming noise. Purchase an inexpensive lavaliere microphone and attach it to the witness. If the unit is omnidirectional, you will get the witness's voice as well as the questioner.

Now that you have invested (or borrowed) a dependable camera, sturdy tripod, and lavaliere microphone, make sure you have three times the amount of tape stock you'll need. I've been asked to shoot one-hour depositions that lasted three and a half hours. When I bring eight hours' worth of tapes, I know that I can last as long as the person talking.

Running your camcorder on AC power is preferred rather than battery power. Unless you are recording on some Caribbean beach (call me if you get this job), the wall outlet will mean fewer breaks in the action. Because this video is legal, you will need to display the date and time during recording. This feature is found on all consumer camcorders, but some prosumer models omit this function. The more ways the authenticity of the video can be documented, the happier the lawyer will be. If your camcorder has no time function, display the footage counter on the screen or use something else that will attest that you have not altered the original tape by stopping and starting the camera. At all costs, keep the camera recording (check for that little red light or "Rec.")

The subject should be framed in a medium shot like shooting a talking head. Camera moves should be left for another project; just capture the image and sound. Different legal people want you to say certain things at the head of the tape but follow their lead with anything along those lines. Your job will be recording and monitoring the video. The lawyer will talk; the subject will answer and begin crying.

There's not much else to discuss about a deposition video. You shoot the entire deposition and hand the tapes over to the lawyer. If you shoot on some format other than VHS, you will have to transfer the footage in order for the legal team to view your end product. Depositions are great ways to pay off your camera equipment investment quickly without much work.

Since this book is about no-budget productions, this is a prime example of just that. It really costs the shooter next to nothing to record a deposition. Of course, you have involved your time, equipment, and tape stock, but what is something like this worth?

Rates vary in different parts of the country, but most small to midsize areas pay in the neighborhood of $100 an hour for deposition videos. When hired by a lawyer, ask what the budget for a deposition shoot is and negotiate from there. If you are shooting in Panavision, charge more. If borrowing Aunt Thelma's 1982 vintage camcorder, charge less but keep its gas tank filled.

Day in the Life

A day in the life video requires much more skill than a deposition. Again, I'm not putting down deposition shooters in any way. It is an exact science and the shooter needs to know what he or she is doing. For the amount of work involved, in my opinion, the pay is quite good.

Following someone around all day long and videotaping everything they do is my best definition of a day in the life video. From the moment your star gets up until the moment he or she retires at night, you must show what happens in a typical day. Most day-in-the-life shootees have been in an accident or are disabled in some way that makes their typical day unique to others. No one would want to follow me around and videotape my day; it would be boring. But if someone has difficulty eating, walking, dressing, or talking, that's what makes this type of video special.

It's almost like you are being the great filmmaker Ken Burns and shooting a documentary of someone special. You will record eight hours of footage and then edit it down to a compressed highlights reel. Like *The Osbornes*, a camera will follow your subject's every move. The camera will see what actually happens to him or her during the course of a day.

The shooting of a day in the life will be grueling at best. You will be on the clock and on your feet all day with the camera rarely not running. This requires much more skill because you may be following someone from room to room (exposure changes), indoors and outdoors (color temperature changes), from quiet to noisy environments (constantly changing sound levels), confined spaces to large expanses (rolling focus), in vehicles, and in public. Your camera should have image stabilization and be comfortable to hold for long periods of time.

The most recent day-in-the-life video I shot over an eight-hour day never allowed me to eat, rest, or do much of anything else. Battery and tape changes kept me occupied when my subject wasn't doing something camera-worthy. Her kids were constantly mugging at the camera and asking if it was on (editing), her dog wanted to be my best friend by smelling me (more editing), and I was asked "What should I do next?" to which I replied, "Whatever you normally do." In these cases, use the best camera you can afford, three-chip models being preferable because you will be editing later (not within the camera). Do not leave all the camera controls on automatic. They have no clue what you are looking for and will be wrong 90 percent of the time. It does make the shooting easier but makes you look less professional.

When you leave at the end of the day with tons of footage, watch everything you shot in the hopes of finding what's good and what may be omitted. To keep the budget as lean as possible and save wear and tear on your camcorder heads, make a VHS copy of all the footage and create a paper cut. Note every shot you want to keep and record the in and out points on paper.

Unfortunately, the next step is unavoidable and will incur spending a little money: editing. No legal person is going to watch an eight-hour video. A 20- to 30-minute program makes them much happier. A *nonlinear editing* (NLE) system is also the only way to fly in circumstances like this. Shots may be added, removed, or rearranged at will without spending hours. Inexpensive NLE systems like Adobe Premiere, Avid DV Express, and Apple Final Cut Pro enable anyone to edit with only spending around $1,000. Of course, all these systems are computer based, so that may be another expense. Figure 1-6 displays an editor working at his NLE system.

Once the video has been recorded, an audio narration track is the next step. Just watching someone all day is part of the process, but the reason why they do something, if it causes them pain, or how they do it differently needs to be explained. Exposition in a movie is when you find out everything you need to know about a character, their history, why they got where they did, and what drives them. You also need this exposition in a day-in-the-life video.

Sit the person down, ask him or her questions, and record the responses. This audio track will then be edited to match the video. The old adage is "See Dog, Say Dog." If you are talking about something, show it. This stage will be the most time consuming because your questions must be edited out,

Figure 1-6
NLE editing systems

mistakes and glitches trimmed, all with correct placement in the video. The only difference between a day in the life and a real documentary is the lack of music. Your program should not entertain, but inform.

With day-in-the-life videographers making as much as $1,000 a finished minute, this could also be quite lucrative. Your expenses are the camera and batteries, tape stock, and your time editing (whether you do it yourself or hire someone). Once again, not a lot is invested to do a day in the life because no scripts are written and no lights are setup.

Mock Trial

The mock trial sounds like something involving a bird and a lot of silly faces and laughing. In reality, a mock trial is the lawyer's first attempt to present a case to a jury, in this case the public. Although I never had mock trials in school, I knew others who were always pretending to be on a jury (I hoped they stopped when they had jury duty). Several people are chosen at random and asked to listen to a case and determine the outcome. This gives the lawyer an idea if they may win a case when tried in a court. When recording a mock trial, two cameras usually cover the group of people, so any grandiose statement may be recorded.

The cameras should be of high enough quality to capture facial expressions, awards of $15,000,000, and hold up to editing. A three- or four-hour mock trial will be edited down so just the important facets are left to be presented to the opposite side. The good lawyer will say to the bad lawyer, "We presented the case to a mock jury and here's what they said." Of course, all the negative material has been edited out. When the bad lawyer sees this edited video, the bum will immediately settle out of court because the case has been lost.

The budget for a mock trial involves getting two cameras that don't have to be identical. In order to save some money, operate camera one yourself and station camera two on a wide shot; zoom in for close ups when people are talking. The second camera is unoccupied, shooting a wide shot, and saving you money.

As in a day-in-the-life video, transfer all footage to VHS and determine what is good and what is not. Present a list to the lawyer and say, "Here's all the good stuff they said," and let he or she decide what stays and what goes. This will save you time in editing. You may think that something was a great statement and spend an hour cutting out all the extraneous garbage only to learn that the lawyer can't use it. Save yourself the aggravation and cost, and allow them to choose what stays or goes in the paper cut phase. The costs involved here: two cameras, tape stock, and editing.

The recording of the mock trial is usually charged as a flat fee (your time, equipment, setup, and travel). The editing (paper cut and final cut) should be charged hourly. No lawyer will pay a dollar per finished minute if the mock trial turns out to be a bust. Determine a fair hourly rate and budget that into the mix. I've spent 3 hours shooting and over 40 hours editing because I have 6 hours of tape that must be logged. I always figure eight times the amount of footage should be spent in editing:

$$\textit{Hours of footage} \times 8 = \textit{Hours of editing}$$

Depending on your experience and comfort level, that ratio may be more or less. Just remember that the lawyer is hiring you for your video skills, not your legal skills. They do not expect you to know what are "good legal words." Just do what you know best, and leave the legal to the trained minds (or the lawyers).

Legal Documentary

Each legal video mentioned has involved increasingly more effort and cost to create. I have saved the most expensive and time consuming for last. The legal documentary differs only slightly from a PBS or Discovery Channel classic. Most documentaries of the legal genre have people recollecting someone living or departed. Each person tells how an individual affected his or her life and about his or her relationship with that individual. The interviewees are lit professionally, shot in a *20/20, Entertainment Tonight*, or *Dateline* style (over the shoulder with the subject looking off camera).

If the documentary involves someone who has passed away, the project may take on a more somber note with much more editing. I shot a documentary involving a wrongful death of a 15-year-old girl. I interviewed her parents and sister, collected her photos and home videos, and edited everything cut to soft, flowing music. This was a challenge for me because I had to present all the facts in the video while still describing her life through still and moving images. The interview process only lasted a few hours, but the scanning of photos, digitizing and capturing three different tape stocks, creating montages, selecting music and graphics took almost two weeks.

This particular documentary was one of my best projects because I took the time (and spent the money) needed to get the job done. When working with any client, determine the budget before beginning the work. The few thousand they spend on you and your skills could net them millions. But that does not mean you should spend freely to ring up the bill. Any extra cost incurred should be discussed and approved by your client.

A colleague of mine is always complaining that I should be charging $2,000 a finished minute for my legal video work. A recently completed 40-minute documentary should have netted me the tidy sum of $80,000 in her eyes. "After all," she said, "your video will get them $800,000! That's only 10 percent! Do you think a 10 percent investment is worth $800,000?" I do, but the lawyer disagreed. The $2,000 per finished minute rule does not always apply. It all depends on your market. Yes, the law firm did win the case and make $800,000, but $100 an hour for my time netted me $5,000. That still pays the bills. I spoke to the lawyer about what I should have charged. He said they would not have spent that kind of money to win the case and I would have been left with nothing.

My point is to feel out the client and see what they are willing to spend. If they can afford $2,000 a finished minute, great; you'll make a tidy profit. But if your overhead enables you to do a job for a few thousand, it's still better than earning nothing.

Never overlook the legal field. Your costs are much lower than most fields and this is an area where video stands out. I know two gentlemen in this business that do nothing but legal videos and earn in the six figures yearly. They have to die sometime and I'll be the one shooting that video.

The Most Uninteresting Video: Documentation Video

Almost every video made is of interest to someone. This documentation category has to be my least favorite. Everyone has opinions; that's why they're free (since this is a no-budget book, that means free so I may state my opinions). You can always try to be creative, but when you are just hired to document something for posterity, creativity flies out of the window. But because it involves the word "video," we must at least mention its existence.

Unlike the documentary, the documentation video may be legal, involving insurance, for real estate purposes, police related, medical, or for reality TV. The documentation means you simply capture the occurrence or document it, which Webster's defines as "to put down on paper." Since it was an old dictionary, substitute video for paper. Breaking each category down into what's involved, let's look at how to do these videos.

Legal Video

As recently mentioned, you can find a lot of video work through law firms. Video is a real and instantaneous medium (unlike the lady who tried to contact your Aunt Sophie at the last séance). Legal minds use video often to prove that something exists. If Knuckles was hurt in a car accident, you may be hired to prove his neck brace is real. By documenting his dependence on the brace, the world will know he isn't faking it.

Some cultures believe that photographing or videotaping the dead needs to be done as proof. I have been hired to videotape a family standing around a coffin with the recently departed. I believed it rather stupid to use a motion video camera where all but one was standing motionless and smiling. But they wanted a video to send back to Japan that had the whole family together one last time. The budget for this is extremely cheap. I shot five minutes of the funeral and gave them the tape. A flat hourly rate was barely enough to make me wear a cotton suit on a steamy July afternoon.

A documentation video has a subbranch in the legal and insurance arena called the reenactment or reconstruction video. This is a misnomer because very little of the reenactment is video, most being computer generated. In an accident because no cameras may have been present to document the actual crash, events need to be reconstructed so the legal and insurance minds may see what really happened.

Once I was hired to trudge out to the scene of an accident and videotape the view from every angle. That video was then married with computer animation to make the fateful accident happen all over again. I recorded the skid marks, I shot at the exact time of day as the accident to catch the angle of the sun, and I tried to get the *point of view* (POV) of each driver.

Computer animators then built the cars and the incident reoccurs in 2D. You've seen this done on high-end news programs and *Raise the _____* specials. The real money in this area is with the computer work. The poor sap shooting the video is paid less than dirt, but it is a good way to get your feet wet in the business. Contact insurance agencies or legal firms to offer your expertise. But before you offer it, you should have some. Do the type of video you like best, practicing your skills before you demand cash.

Dead people also like to be on videotape when their wills are read. Even though it doesn't take much talent, I've been contracted to shoot someone, let me rephrase that, tape someone, sitting in a chair reciting his last will and testament. Like a deposition, the camera doesn't move and when the soon-to-be-departed is finished, you hand him the tape and it sits in his safe deposit box until he moves on (not to Toledo).

Basically, any video that requires you to record proof of something falls under this glorious heading. As a side note, if this is something you were born to do and you advertised this as your specialty, you may never get any other type of video work. The video world is a small, tightly knit community. Just like Fred Gwynne, (aka Herman Munster), people seeing you do something well will believe that's all you can do. Be versatile and if you must create a list, list everything you do, not just a few things.

Insurance Video

At least in this area you are allowed to move around a little more when documenting. For insurance purposes, everyone should videotape the interior of their homes. This is a fabulous way to inventory all the neat stuff that your wife, husband, or mom wants you to throw out.

With a wide angle lens attached to your camcorder, wander around the home in question and record every object in each room. With a little forethought, write down the serial numbers and speak them as you shoot. I don't mean bark them in dog language, but speak clearly and you will have a visual and aural record of lots of substance.

The chargeable fee for this needs to be low because any homeowner or renter can get a camera for a few hundred dollars and do it themselves. Once again, this is another zero-budget production and your only investment is the tape and your time.

Real Estate Video

This market has dwindled in some areas and grown in others. People today would rather shop for their six-figure home on the Internet than actually drive over and see it. With the magic of virtual tours, the browser can view the home or office from any angle without leaving his or her home.

I did a lot of still photography for realtors in the olden days, but most agents carry digital cameras in their suit pockets and accomplish the same feat themselves. However, few carry video cameras because that involves a little more work. Because of their laziness, we can get extra work.

Shooting a virtual tour of a home is as easy as it sounds if you are hired just to shoot. Some have done this in the past with a series of digital still images, but motion video (analog or digital) is sharper and enables more movement. If hired as the video person, shoot the exterior, every room, from every vantage point. On a computer, the images will be captured and a branching path will be created so the viewer can see what they want to and when. The programming is somewhat difficult but well within the grasp of someone in that field. I just hand over the digital tape and they do the hard work. It's now possible to do 360-degree shots of every room. Remember you are a professional if you are paid for what you do.

Police Related Video

Even if your relatives aren't cops, police videos are very popular. Tune into any police reality show and see that video is everywhere. I don't think most readers will be involved in the craziest chases or shootouts, but I too have spent the day in a police car (sometimes actually videotaping and not handcuffed) and videotaped their every move.

Luckily, my days were boring and nothing happened. But every time the police stopped the car (hundreds of times) I had to get out and follow what they did. This seemed just like a day-in-the-life video, only on steroids.

Getting in and out of the car a thousand times is enough to make anyone a tired camper. Friends of mine in larger cities have to wear bulletproof vests while working on these events, but with my luck I would be shot in the head on a local police fundraiser. If put in a situation like this, make sure it's really what you want to do.

Because I was shooting for national television (with my local police), my equipment (camera stuff) and my time (all stinking day) were on a professional level. The only expenses incurred besides coffee and donuts 12 times a day were the equipment rental, tapes, and my time. Trying to light, getting a sound check, a focus check, or setting the correct color balance were mostly left to the imagination. Why do the bad guys always run into the darkest corners where the camera records only noise and grain?

Don't use the night-shot feature on your camcorders unless the shot can't be done any other way. Infrared is featureless, has a green cast, and only works three feet from the camera. Professional cameras omit this feature because only amateurs seem to use it.

One other item that rings up the budget with police videos is hiding the faces of the perps (that's cop talk). This involves editing. If you are hired to shoot, ask if you should just turn over the tapes or if you are required to blur the faces of the naughty people. If so, set your price accordingly. Blurring or pixelizing in an NLE system takes more time than with a video switcher.

The other police-related videos are the aftermath of a crime scene or forensic work. I really dislike this type of video because it's too real. The gore, sights, and smells are not for the queasy. But sometimes you are hired to videotape something exactly as it occurs before the police or coroner arrives. Don't touch anything; just do your job and leave (before dinner if you are trying to lose weight).

As part of the police-related documentation video I did, I was hired to record a police and fire exercise for the *Federal Emergency Management Agency* (FEMA). On a hot, late August afternoon in 2001, I videotaped a staged terrorist attack on a small park. The edited tape would help train emergency management people how to best handle a terrorist attack. Using 500 extras as victims, the terrorists attacked a train load of people in the park. Less than one month later September 11th happened. FEMA had no clue that this training would be so timely, but you saw the immediacy of video that day.

Medical Video

Before agreeing to this lucrative type of work, make sure you know what's involved. I am extremely uncomfortable shooting surgery on people's eyeballs. I've taped knee surgery and bone setting, and had my wife remove a splinter from my finger without me passing out. When the laser beam hits the cornea, I'm scaling the walls to get out.

I've been told just to look away, but that makes it difficult to frame and focus. That's why the operating rooms are so cold; if they were warmer you would pass out more quickly. If you can distance yourself from the surgery (like across the street) and shoot it like it's any other job, life is grand. I know this is something I cannot do, so I back out slowly. If you have no problem with this, you will be paid well for your time and trouble. Medical clients pay two or three times what most get for an average shoot. That may be because it's not something that can be repeated.

Keep in mind when you are shooting a documentation video, you rarely get a chance for take two. This is the only time where you might, and I say might, use autofocus and exposure. With everything happening so quickly and everyone trying to revive me all the time, any time saved is important. Medical surgery doesn't have to be repulsive, as in Figure 1-7. At least this person wasn't on his knees.

With any documentation video, know what you are getting into before you accept the job. Even if the pay is fantastic, if you spend all your time sick, uncomfortable, or unconscious, follow Nancy Reagan's advice and "just say no."

I was once hired to shoot an autopsy. Don't ask me where I get these jobs; I'm still trying to find that out myself. The camera was set up on a tripod as the examiner did his deed and spoke into the microphone. There was no reason for me to be there. The camera could have been rented and I could have been enjoying a thrilling game of Tiddly Winks. For some reason, they wanted another breathing human in the room (at least there was no blood).

Everything was as boring as can be until the examiner put a tray of tools on the cadaver's stomach. With four pounds of pressure on his stomach, the trapped air escaped as he slowly sat up and gasped, "Ahhhhhh!" I don't know what it takes to scare an examiner, but I know what it takes me to wet my shorts.

Figure 1-7
Medical video

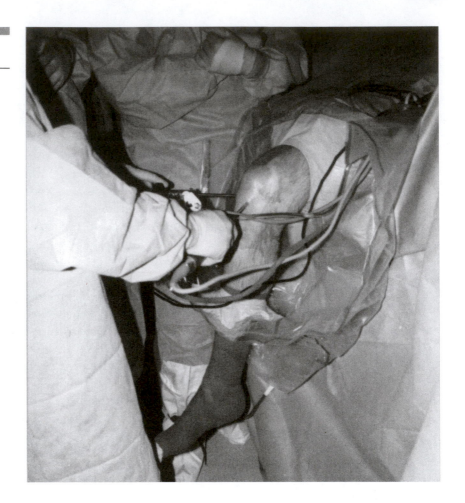

Reality TV

This last grouping has got to be the hottest genre of video—this week. With TV shows like *Survivor, Fear Factor, Mole, The Osbornes*, and all of MTV's various installments, people want to know what is happening . . . now.

On the lower end of the scale, I was hired by the Learning Channel to follow a pregnant woman around and videotape her child's birth. They had tailed this lady through her entire pregnancy and my 12-hour shift occurred when she was already 2 weeks past due. Like a Butterball Turkey, she was ready to pop any time.

Being very young during my own birth, this was the first time I was witnessing a stranger coming into the world as seen through the viewfinder. In health class, we had cave paintings to document the birthing process. I would now be in the hospital seeing the real thing. Seeing is the easy part; hearing and smelling it is quite different.

As I said earlier, know what you are getting into. In reality, you are exposed (a film term) to some pleasant as well as some unpleasant things. Your job is to record the event impartially, without emotion or feeling. You are there to make sure everything is captured on video.

In film school, my instructor told us that if hired as news photographers (using motion picture film because video wasn't available) our job was to film the event. If we were the only people on the scene and someone was dying and needed our help, our role was to shoot first. After we got the shot, then maybe we could help. I thought that was heartless, but that is in fact what you are paid to do. If you help the person, there may be no story and no money.

With video cameras on the heads of missiles, we saw immediacy during the Gulf War. The split second something happens we can see it live on TV. You may be a local stringer for a news service. If some event occurs close to where you live, you may be called out to make video history.

No matter what aspect of video you chose, make it your life's work and do the best you can. Others are depending on your skills to see these events. Make them proud.

Everything mentioned here costs very little, if nothing, to shoot. Documentation video is no-budget production at its finest. I wish we could only get people to believe that it's not a "no-interest" production.

In the Beginning: Preproduction

Getting Contacts

Being independently wealthy has always been the way to fly, but since most do not have unlimited funds at their disposal, we need to see how others do. In the world of video, knowing other people or having contacts allows you to spend the rest of your days doing what you love to do.

In this very competitive business, many others are vying for the same opportunity. Knowing people is the easiest way to get involved in making videos. My first bit of advice to the neophyte video auteur is to do as much as you can. Make videos of anything and everything. Always have the camera in your hand, up to your eye, or with you when you travel. The best way of getting good at something is doing it.

Once people have seen you with the camera and know you two are inseparable, the next step is to remind them exactly what you do. Tell everyone you know (and some you don't) that you make videos. Telling people is a great icebreaker and most will say, "Oh, really?" or "That's interesting," or other trivial phrases unless they have an immediate need for your services.

This word-of-mouth approach gets your name out in the business. Your objective is to get contacts, but you really have no idea who a contact may be. In actuality, anyone you meet or speak to for any reason is a potential contact. More than three-fourths of my freelance video work comes from contacts I don't even know I've made. I've told someone in passing what I do for a living while seated on a plane, someone has seen me shooting something in a public place, or I've been referred to them by a friend (I only have two friends, but they work really hard).

So how do you establish contacts in the first place? Throughout this book, I'll be compiling lists and this may be the first. Five ways exist for establishing contacts: advertising, referrals, business cards, word of mouth, and seeing is believing.

Advertising: Always the Best Medicine

The best way to sell anything (yourself included) is to let the world know exactly what you want them to know—advertise. Since everyone is media savvy these days, we are aware how advertising works. You've all seen the commercials that left you chuckling afterwards, but do you remember the product advertised or were you more impressed with the message? If you enjoyed the spot and still remembered what was being pitched, then the advertising was effective. You need the same approach; leave a lasting impression but also sell yourself so others will remember you.

Advertising takes on many forms: written, aural, and visual. The easiest and most widely used form of advertising is the written word or ad. The phonebook with its Yellow Pages lists businesses in single-line or in various-sized ads. When you are looking for a particular service, which one do you usually contact, the one with the company name and phone number or the boxed advertisement? Most will call the larger advertisement because it catches their eye, the company is probably larger and knows more (because they spent more money for the big ad), and more information about the company is listed without you having to ask them numerous questions.

While we're on the topic of phonebook advertisements, we'll discuss the best way to get local contacts. Every company open to the public is listed in the phonebook. If you look under the Video heading, you'll see all your competition. When checking out the competition, see what works for them in their advertisements. If just starting out in this business, offer services to the other video production companies in your area. They have already spent money advertising, marketing perspective clients, and probably already have established clientele. They may need to gear up when a shoot occurs and need freelance help.

A subheading under advertising would be freelance work. While working for someone else in this business, you are actually selling yourself subtly. You shouldn't be broadcasting to everyone on the set how great you are, but rather do the best you can in your specific role and others will notice. I used to like wearing a T-shirt that listed my chosen profession so the world would know what I did on a particular shoot, but that backfired when I had the word "Grip" printed on my T-shirt (I did, however, make a lot of new friends that day).

Freelancing also relieves a lot of the headaches working totally on your own creates. You are now being paid for what you do and don't have to try to drum up new work that day (don't ever do that while being paid by someone else. It's their dime so work your hardest for them). Clients are always complaining about something and being part of a freelance crew, you just do what you are told and you won't have to bother with most of the unpleasantness.

When I need freelance help on my shoots, I contact others when needed and they do the same with me. Basically, I use my contacts. As a producer, it's my job to solve problems and find the people or services my client needs. Being in this business numerous years, my list of contacts has grown with each project. Whenever I need someone, I use my contacts to fill that need. After your first freelance shoot, you will have at least 10 new contacts. Like a pyramid or legal chain letter scheme, if you get contact information on everyone in the crew, potentially you could get 10 new jobs (one from each of the contacts) on your first go-round. Each of those 10, if they like you, will give your name to others, and so on. If you land your own gig and need help, you may contact someone from your growing list.

This list should just be used to get freelance work on crews. If you are working for Martha's Buttered Bacon Products as a freelancer, don't contact Martha and try to land your own freelance work if the director hired you as a grip. This is called stealing and you won't be working freelance long if you resort to this.

Now that you have a new listing of contacts, talk to them occasionally to keep the list active. It's embarrassing when I call a contact after a four-year span with potential work only to find they moved two years earlier. This is now a dead contact. Even if you just send an email every few months, let people know what you are doing, if you are available for work, and ask what they are doing. Everyone loves being asked what they are currently doing and maybe you could fit into their plans. It never hurts to talk to your contacts; once you stop, they may cease being contacts.

The Web is another great form of advertising or getting contacts. A Web site called www.mandy.com based in London offers job and crew listings all over the world. If you need anyone, doing any job, in any place, this is a one-stop shop. I can now tell people that I have video contacts all over the world, and I do. I once needed video crews in eight different countries around the world. By spending a few hours on Mandy's, I had crews lined up in each place. An example of mandy.com is shown in Figure 2-1.

In addition, by listing your pertinent information on a site (like Mandy), the world now has you as a contact. I really don't know how we got along without the Internet as a listing reference.

The best place to start with Internet advertising is to set up a personal web site. I will go into more detail in Chapter 10, "Streaming Video on the Web," on how to best sell yourself on the Internet, but do use it. Your source of new contacts is almost limitless.

Other forms of written advertisements may or may not be effective in getting contacts. The listing on the public Laundromat bulletin board may get you a few names (I won't discuss this further). Postings on telephone poles and flyers in barber shop windows don't have the pull or charm they once did.

Referrals

The only way to get referrals from contacts is to have pleased someone with your past performance. If your last freelance employer or client liked the job

Figure 2-1
Mandy.com

you did for them, they will, without hesitation, refer you to another colleague. My grandfather, Irving S. Gloman, once said that, "There is no such thing as a bad contact." Everyone you know can potentially get you more work. If you do a poor job, you still have a contact, but they will not refer you to others. I once ticked off a client who thought I should be mopping the floor (because of my lack of experience) and whoever hired me should be holding the bucket. This individual was still a contact, but not one I spoke to frequently.

Most of my freelance work is from referrals or from the Web. Local work, once you establish a group of contacts, will grow exponentially. People get sick, two shoots happen on the same day, or a large project wanders into town. Your name may be referred to pick up the video slack.

A tremendous video referral service is your state's Film or Video Commission. Every state in the union has a Film or Video Commission whose sole job is to locate and refer people, crews, and services to those seeking just that. When someone comes to Pennsylvania from either in or out of state, unless they are in the video business, they don't know who to contact. Your state's video commission handbook will list your name and pertinent information (usually free) as well as others throughout the state.

The Commission cannot pick one person over another for a job, but they will refer the client to their listing. If you are not on your state's list, contact the governor's office and make it happen. Believe me, it's more fun than trying to get a pardon.

Parents, friends, and relatives also will refer you whether they know it or not. Face it, video is a glamorous business to most and it fascinates them that someone can make money having fun. Just look how much attention you receive when you are standing next to a camera.

I shot a commercial for an upscale jewelry store where the owner and his wife both dressed in diamonds and formal evening wear and rode down the sidewalk in a vintage Harley-Davidson. The sidewalk was roped off and lights, sound equipment, and video crew were everywhere. When a tourist walked by all this activity, who do you think he stopped and spoke to? Not the guy on the motorcycle, but the one standing next to the camera, because that's the part that was real and obtainable to him. If I sat on the chopper, no one would have given me the time of day (because I was wearing a Rolex).

Parents are a great help too. I've received video work from my father through his unbiased referral. My sister's business has also netted me income. Use your family and friends (only for good). If they know what you do, they can sell for you.

Business Cards

Everyone wants to have these! From your first professionally printed card when you graduated from high school to the computer printed tear-offs you give to colleagues, business cards are an old-fashioned but still effective means of letting people know about your talents. You can also cram a lot of important information as well as artwork on that little piece of cardboard.

Business-card video CDs have established themselves as the only card that, once used, will not be discarded. I know you have a million business cards from all the people you've met in your sordid life. All these cards have vanished, evaporated, or have been used to remove gunk from under nasty places. But no one has ever thrown out (intentionally) a CD business card. This is more than just a novelty. This is a great advertising tool. A recent study proved that if a catalog were mailed on paper or VHS cassette, 98 percent of the people would view the videotape (no matter what the content), while 85 percent would throw out the paper catalog without looking at it. The same principle applies to a video-based business card. This novelty is an effective tool to sell yourself and show your demo reel in the process.

Speaking of demo reels, this is one business where your talent alone does not always land you the job. I've produced, written, and shot almost 800 commercials and every new client wants to see examples of my work. I still need to update my demo reel often. Any position in the video crew arena or any type of video production needs a demo reel. It's unfortunate that just selling yourself isn't enough. The selling will get you in front of the right people, but the reel will get you the job.

As your skills increase, update your reel to reflect your improvement. Even getting that first job demands a reel. The Catch-22 of how to get a job without a reel if you need a reel to get a job is easily overcome. Go out and shoot a demo reel. People are more concerned about what you can do rather than who you did it for. Even with phony clients (if you did not do the work for them, do not list that you did), a compilation of your best work will get you noticed. This is a great place to use all the video footage you shot honing your skills. Put it together in a five-minute or less package to highlight what you do best. You now have a demo reel.

Word of Mouth, or Watch What You Say

I firmly believe that I get more freelance work from talking to people (people always talk about me) than any other avenue (and I live on a crowded street). If you enjoy what you do, why not tell people? Once again, don't brag about your skills, but work your love, or strong like, of video into the conversation. The key here is to be subtle. Don't keep turning the conversation to you or change the subject. If people know what you do, word of mouth will happen by itself. Make yourself visible and soon you'll be seen better with your new contacts.

Seeing is Believing

There's nothing like hanging out of a cherry picker shooting the de-icing of a Federal Express 727 aircraft's tail section in a snowstorm to get people's attention. If that's not enough, try dangling from a 1949 Bell Helicopter over 100,000 people at a concert and try to be inconspicuous. When people notice you shooting video (it doesn't have to be as dramatic as the last two examples), their brains will put two and two together.

Only one person in that concert crowd may have needed an aerial footage shot. If they had that need, they may get more information about the "guy in the helicopter." Or they may want my name for the lawsuit. What I'm trying to say is to get noticed. Do your job (without attracting attention) and you will attract attention.

Every school I've ever shot in (videotaped), the kids are attracted to a camera like a magnet. I won't get much work from the kids, but the schools remember when they need video work. Shoot parades, pageants, or other outdoor events and people will notice you. I know this may sound like an ego trip, but when people see you have a camera and shoot things, it will lead to more new contacts.

Making yourself available is the last bit of advice I will discuss about getting contacts. If you are always reachable either by phone, pager, or email, people will continue contacting you. If you don't return their inquiry quickly, they will move onto someone else. They have an immediate need and want it filled quickly. No matter how boring, mundane, or uninterested the job lead may be, hear the people out and refer them to someone else if you can't or won't do it. They may have more work in the future or know of others who need your services. Never just say, "I'm not interested." If you do, you just lost a contact (and you know how hard it is to find a lost contact). Save contacts, trade them for valuable prizes, and touch base with them often and they will pay off.

It's always good practice, once you obtain a lead or contact, to find out where that particular lead originated. How did they get your name? This approach lets you know which methods are being effective. Was it the Yellow Page ad, the business card, an Internet listing, a referral from a friend, or did someone see you working? Once you determine how they got your name, reward or thank that person if possible for sending work your way.

Unless you have a specific area in video where you like to specialize (corporate videos, event videography, or commercials), try your hand at all facets of the field until you find something that you love to do. Contacts have a way of finding you; maybe that's why people prefer them over glasses.

Approaching New Clients on Your Own

Now that you have a long list of contacts, have decided which type of video interests you most, and are ready to start earning money, it's time to start approaching clients. Often you may have to sneak up on them to get their attention, but let's try the more subtle approach.

The key is to leave no stone unturned when looking for new clients. You will meet all types in the client world, but you need to know where to find them. Some of the avenues for hunting new clients are the phonebook, the Internet, newspapers, job fairs/trade shows, and cold calling.

Letting Your Fingers Do the Walking

The easiest locator is the phonebook. I mentioned this approach in Chapter 1, "What Kind of Videos Can I Make?" when looking for contacts. Every listing in the yellow section is a business of some type that serves the public. You can start with the letter A, but that wastes a lot of time.

The size of the advertisement (as mentioned earlier) determines the amount of money the particular patron spent. The bigger the ad, the more it cost. A large, full-page ad doesn't mean that the company is larger or better than one that receives a one-line listing; it simply means the large-ad people are paying more to get noticed. That may be a good or bad thing. Check out the Yellow Pages listings for Attorneys. Every lawyer has his or her own full-page ad. If you happen to run across a one-line attorney listing; they are new in the business.

Look in the Yellow Pages and call only the listings that interest you most. I'll get into an example of how this works later.

Come into My Chamber

Another new client resource is your local Chamber of Commerce. The Chamber lists every business in your area and knows a lot about them. Before you call or visit a perspective new client, learn all you can about them from the Chamber or Internet. This research lets the potential client know you've done your homework. If the company is heading for bankruptcy, they probably won't want a video made (and you'll learn that from the Chamber).

The Chamber will also have a listing of new businesses. This may be the best place to start. A new business won't have a record in the phonebook until the next one is printed and they may need to let the world know of their existence before then. Although almost half of all new companies fail, in the beginning they are willing to spend money (your budget) to get noticed. Follow these leads to track down new businesses.

The World Wide Web

The Internet also will list most companies with two or more people. I don't believe that using the Internet alone to find business works that often, but it is a great tool in conjunction with another resource. Use the Web as a second link once you have located a company. Look at the About Us, Information, or History sections to gather your background information. The more homework you do, the better prepared and knowledgeable you appear to the client. The Web is also a worldwide resource. Begin locally and then you may expand to the rest of the planet.

You can never be overprepared when meeting a new client. Use the Internet for good and stay away from the naughty sites.

All the News That's Fit to Print

The daily newspaper, other than the classifieds, lists numerous potential clients. Every week the newspaper spotlights a new company and goes into detail why they exist, what they do, and so on. I usually check the Sunday paper because it's larger and has a monthly section on businesses. Through one three-inch ad, the experienced client hunter can determine if they should be approached for video.

This head hunter looks for words like expanding, new hirings, new office opening, large expenditure of new equipment, wanting a larger presence in the community, or overstocked inventory. Key words like this mean they need to increase sales to pay for all they've just done. Future chapters will discuss whether they need commercials, how-to's, or documentaries. For now just focus on getting the clients.

All the World's a Job Fair

Job fairs are a fantastic place that's free and full of company representatives (all wearing suits). You may easily wander up and down the isles and see what each company has to offer. Some have a video playing, which means they've done business in the past. Watch the video and see what works and what doesn't. The company representatives will see that you are interested but not the reason why.

If you care to pursue this, ask them if their video has been effective. They may have had to throw one together just for this trade show. If their video client isn't married or in a loving relationship with the vendor and was hired by bidding, ask if you can bid on their next video. The best thing is just to keep your eyes and ears open. Every contact could lead you to a new client. If you are looking for work, it will find you. Just make sure you are ready at any time to propose a video.

Boy, It's Cold in Here

Cold calling, as the name implies, is difficult at best. You've heard about a particular company from some avenue and now you're checking them out. Whether you walk in and ask for the person's name or call them on the phone, you are still doing things the hard way.

If you have friends that own businesses or work for a living, try them first for potential videos. They know you and your abilities, and it's a good place to get your feet wet. Start small and ask them if a video would make them happier campers.

Is It Safe?

Few businesses are recession proof, but health care is one of them. It doesn't matter what shape the economy may be in, hospitals need to carry on. That's good news for video people. When most businesses are slow, I've always found work with health care. They may have funding and budget concerns, but the economy will not affect that.

The Big Example

Let's look at a specific example on how to gain a new client. This is a true story; only the facts have been changed.

First of all, decide what type of video you'd like to do. The sky's really the limit. I thought a good place to start would be food commercials. I don't know why I began there, but as you read on you'll see that I'm kind of strange . . . unique anyway.

I got the local Yellow Pages (my favorite source) and looked under Restaurants. The restaurants that just had a one-line listing I didn't even call. The ones with the ads were my victims.

Lancaster is a pretty good-sized city and I called over 50 places (each one had an ad in the Yellow Pages). It's important to talk to the right person whenever you call a prospective client. The person who answers the phone, although pleasant, isn't usually the person that makes the advertising decisions. Ask to speak to the manager or owner. Some people are a little nervous when a person calls and asks to speak to the manager; they usually ask if they can help you or what this call is in regard to. You aren't selling nuclear weapons and you have nothing to hide. Be courteous to them and explain your case. Sometimes this can save you a lot of time if the business is so small they spend no money on advertising or if the manager is in prison for tax evasion.

Out of 50 calls if you can set up 5 meetings, you're fortunate. The calling is the most difficult part. Ask anyone in sales. You have no idea what you're going to run into when you make a cold call; it's tough. But it's better to let your fingers do the walking instead of your feet (it's an old slogan, but I changed it).

I won't go into the obvious as how to dress and present yourself at this first face-to-face meeting. Great books on sales techniques can be found out there and they will be more helpful. If you really want to make videos, your enthusiasm is your best tool.

It's best not to go in there unannounced. You may end up sitting in a chair staring at the receptionist for three hours (it depends on what they look like). If you make an appointment, be prompt.

When you get there and talk to the manager or owner, find out what type of spot they would like: humorous, serious, informative, hard or soft sell, or price related. They will usually have no clue what they want. Explain what the different types mean and how you'll approach them. Also ask them if they have a video resource already. If they do, it's best not to step on toes and muscle into someone else's turf. Solicit if they are satisfied with the vendor; if they are, move on and possibly see if that company needs freelance help (not to steal the account away from them). If they aren't happy, inquire what it would take to make them happy.

I'll give you an example of what happened to me (get used to it; I'll be doing this a lot). After calling a ton of places and only having a few positive leads, I was frustrated. One evening I went to a new restaurant (it wasn't in the phonebook yet) with some friends. This restaurant specialized in Cajun food. Figure 2-2 shows a behind-the-scenes setup for a food shoot.

Since the establishment was only a week old, it was empty. We were the only customers in the place. After looking at the menu, I realized that this place was unique. They offered food items like alligator, turtle, conch, and snapper, and it was presented in a unique hand-sculpted, edible container. This got my mind going. With food this visual and different, I could easily see a commercial.

Figure 2-2
Food setup for
shooting

Since the restaurant was new and understaffed, it took a long time for the chef to create his masterpieces. But the food actually tasted good. After the meal we sat around a long time waiting for the check. I used this time to my best advantage. While waiting for the check, I talked to the owner.

I used a truthful sales approach (does that really exist?) and told her how much I enjoyed the meal and its presentation. Immediately, the owner liked me because I started off with something positive (not because I was flirting —I wasn't; I'm married). If I had said that the alligator was good but tough, that might have put her on guard.

She opened up to me and said how difficult the alligator was to prepare, how the recipe had been in the family for generations, and lots of other info that I could use in a commercial. I mentioned it was too bad that no one really knew about this great eating spot. I told them they should advertised and get the word out.

She agreed but didn't know how to go about it. That's where I came in. "I happen to produce commercials. Would you like some help in creating one?" Her eyes lit up. Yes, she was interested, but how could she afford it?

That "cold call" with the owner had gotten me further than a phone call would have. I had sampled the food, she had a need, and I was in the right place at the right time. Use this sort of situation to its best advantage. Try new businesses. Check them out as a customer and if you believe there may be a need for a video, present it to them. Always have a business card with you because you'll never know where a possible lead may pop up.

I met with the owner a few days later for a meeting. It's tacky to expect the owner to meet with you right on the spot; both of you aren't really prepared. I asked the owner if she had a slogan; people like slogans. Because they were new, they didn't. Brainstorming I offered some stupid ones: "An alligator you like to eat," "Cajun food: the spice of this life," and "Indigestion is on us."

Where I grew up, a septic tank cleaning firm had a great slogan: "We're number one in the number two business." It sounds kind of disgusting yet is very effective. I saw that slogan 20 years ago and I'll never forget it.

A slogan is just one way to get the viewer to remember the spot, but it wasn't working with the Cajun restaurant. It had to be something visual for this place. Because every piece of food was creatively prepared so it looked great, the spot had to feature this. This and the type of food they served set the restaurant apart from the others. The owner agreed.

Now that you have a concept, how do you pull it off effectively? As you read on, you'll figure out that I like to use humor in videos. Humor would play quite well with my approach.

The only way kids will eat anything is if it looks good. If it's good for them, forget it. So how do you get a kid to eat an alligator? "Come on, Junior. We're not leaving until you finish all of your alligator! People are starving in (insert country), they would be glad to have your alligator."

The spot would open with a full restaurant (makes it look as if it's popular). Lots of kids are there too (good, it's a family place; I don't have to get a sitter). Everybody's getting exactly what they want and every type of food they have is shown being served (they have a large selection and it looks appealing). A young man is trying to impress his date by eating spicy Cajun food (all types can eat here, wimpy stomachs and cast-iron ones).

Through little scenarios at each table, we showed that the restaurant has all types of customers, the food looks good, there's something everybody likes to eat, and it's a fun place to spend the evening.

The cost for this spot was very low because I wrote, produced, and shot the commercial. I rented the equipment and purchased the airtime at a local TV station. I had fun creating the spot, the client got an effective piece showcasing their restaurant, and the world (Lancaster) was able to know this place actually existed.

Now the restaurant is in a new and bigger location, has quite a following in the community, and is extremely busy. All this because of the commercial I created for them. At least I like to believe that.

There is only one wrong place to meet new clients: a funeral. Any other situation is ripe and should be used to its fullest.

How to Get Ideas for Videos

As soon as people find out that you have anything to do with television (besides knowing how to turn it on), they immediately think that you are extremely creative and therefore full of ideas. Although that may be the case, it isn't always so.

One of the most difficult parts of creating videos is coming up with new, fresh ideas that will set the world on fire without overtaxing your brain. Television has been around for a long time and videos (once called films) have been around longer. A lot of the old standard ideas have already been used, and most of the new ideas out there today are bordering on the bizarre.

Just suppose you are a top-notch video idea person and a client comes up to you and says, "I have money to spend (they never say exactly how much). Come up with some ideas for a corporate video." If you aren't sure what a corporate video entails, read Chapter 4, "The Corporate Video." Now that the money has been dropped in your lap, how and where do you get ideas to keep this person happy?

For every problem, two ways of thinking exist: the good and the evil, the honest and the dishonest, the nice and the not so nice, and so on. The first way is to be creative and come up with an idea or concept, and the other is to borrow it from someone else. For the sake of this book, we will be doing *no* borrowing. The most sincere form of flattery is imitation but not necessarily in this business. Someone came up with a great idea, got paid a lot of money, and probably owns the rights. We will not go this route; it's unfair and if they are bigger than you are, you might get hurt. Besides, you don't want to share a cell with someone named Fingers. Be creative on your own; don't use someone else's ideas.

My ideas for corporate videos only come from original places (yeah, right!). I recommend that you watch other corporate videos (you have seen them before; they explain why a particular business is better than another).

You should watch these not to steal their ideas, but rather to see how others approached the same subject. After watching them, can you improve on the concept? Where did their video go wrong . . . or right? Critiquing is important; you can learn from their successes and mistakes. This is another reason why I got into video. Anyone can critique something. Opinions are free; that's why everyone has one. It's far more difficult to make something than to complain about what someone has done. If you know a better way (and you probably do), do it rather than griping how bad the previous video may have been.

Learn as much as you can about the company you are going to advertise. Read all the literature you can find, check out any print ads if they exist, visit the company, watch what they do, and talk to the employees. In order to best sell a company, you have to know all there is to know about it: What are its strengths and weaknesses?

Brainstorming is another way that can help the flow of ideas. Sit and stare at the company literature and let your mind wander. What comes to mind? What concept keeps popping up? Elaborate on that idea; let it grow. Sometimes it helps to bounce your initial idea off someone else who isn't as close to the project. How do they feel about it? Does it get their attention?

I had to write an infomercial on a package of CD-ROMs that would make you a top marketing executive overnight. The package costs thousands but claimed you would make millions in less than a year. It's difficult at best to write about something you don't believe in. I thought it was a scam, but I forced myself to believe otherwise.

The only information I received in preparation for writing the script was a four-page blurb on their web site. Obviously, it was all propaganda, but nothing else could be learned about this CD set. After spending too much time reading their garbage, I came up with the approach of looking at it from my point of view (a nonbeliever). By turning a skeptic into an advocate, the infomercial was well received and I still got paid.

Sometimes you get ideas when you aren't thinking about how to get ideas. If staring at the product is starting to make you ill, do something else. The project will still be in the back of your mind. Soon something you see or hear will trigger your thought processes again.

Reverse Psychology

Like my "Get Rich Quick" infomercial, try to think of what the corporation is not. Sometimes this reverse type of psychology really works. If you were doing a commercial about fish food, try to imagine you're selling it to kids. Kids couldn't care less about fish food, but if you present it to them on their level, they may go for it. In other words, try pitching the video to someone it isn't intended for. This sounds a little off base, but sometimes if you do things in a roundabout way, you'll come up with a novel approach (or maybe a movie deal).

Look at the competition's videos (other corporations in the same business). Whatever they do, try just the opposite. Sometimes this will give you a fresh approach. How do you see the company being advertised? What facet of the business would make you want to spend money on them? Would a kid pitching the company help? Try to see it from their eyes (only taller). The innocence of children may be helpful (or nauseating). Do you want to be told about the company via an onscreen face, a voice-over narration, a little scenario, or through graphics?

When I have a hard time coming up with an idea, I find the most boring lecture, free discussion, or recital I can attend (try to find a free one; they're cheaper). After sitting there for a short time, my mind wanders and I have a concept. Or sometimes something the lecturer says (through my half-listening ears) sparks my interest. Don't do this in church, while driving, or while your spouse is yelling at you. If you're listening to something or doing something that doesn't interest you, your mind will wander and that's when your subconscious kicks into high gear. Sometimes just turning on the TV and half-watching it, or listening to the radio (but not listening) helps. If your mind is someplace else, hopefully it's in a place where ideas originate. Look at the pictures in catalogues, read the synopses of movies, get stuck in a traffic jam, pet a cat—whatever causes you to think, that's when you'll get your best ideas.

Review: The Nine Most Important Ways to Get Ideas for Videos

The nine ways are as follows:

1. Find out exactly how much creative free reign you have. If the client has something in mind and they don't tell you, you may be way off track.
2. Watch other videos of the same caliber. What worked and what didn't?
3. Don't steal someone else's idea or concept. Be creative and come up with your own. How would you best sell the company's service?
4. Learn as much as you can about the company. If you're going to sell it, you better believe in it and know it pretty well.

(continued)

(continued)

5. Try brainstorming. You'd be surprised what thoughts enter your mind. Be careful using this because your end result is usually far removed from your original concept.
6. Get a fresh perspective. Tell someone else about your idea. Does he get the concept or is he just obtuse?
7. Stop thinking about how to get ideas for a while and think about something unrelated. When you stop concentrating, that's when fresh ideas can find their way in. Keep your ears clean.
8. Go someplace boring or do something you don't have much interest in and let your mind wander.
9. Try different twists or endings to the same idea. Ask the client if humor is okay, storytelling, and so on.

It's hard to tell someone, "This is the best way to get ideas." As soon as you say it works for you, it won't work for them. Coming up with ideas can be done in a thousand ways and I'd tell you every one of them, but I'm running out of space. Don't be afraid to try anything unusual to get you to think (as long as it's legal). That's the key thing to remember in all of this. What gets you to be the most creative and thinking? Go there and do just that. There, you've already solved the problem. That will be $50.

How to Write Low-Budget Productions

It helps when you know you have limited funds and you can write the script with that fact in mind. You may still come up with great concepts, flashy action sequences, and neatly dressed aliens, but the writing should reflect that.

When the budget is miniscule, keeping that concept in the back of your mind when writing allows you to dream in a slightly different way. If you have an action sequence on the Brooklyn Bridge and you need traffic tied up for hours, possibly a little money might be needed. But if you just get a long, establishing shot of the real bridge and shoot the rest of the material on another local bridge in close-up, no one may be the wiser.

No Class Today, There's a Substitute

I like to call this the substitution rule. Whenever possible, substitute a cheaper item in place of the real, more expensive original. In *Silent Running*, Bruce Dern is trying to kill someone by holding a shovel against his throat. When the fight began, a rubber shovel was substituted for the metal one. Unfortunately, as the actors fought, the rubber blade wiggles as they struggle. This time the substitution didn't work.

Hollywood cheats on us all the time. Not to burst any bubbles, but most of the nifty locations in movies are really just sets because it's cheaper to shoot that way. In your videos, show the exterior to establish where the actors are and then cut to a set or other less expensive location. The viewer will still think you are at the original local. Why? Because you just showed it to them. You have great control with your writing, shooting, and editing. Making people believe and see what you want them to is a god-like power. It's your story and your video, and you should be in total control. Just don't let your power slip. The viewer wants to be directed and pulled along; don't disillusion them.

Read My Lips

This "fooling" of the audience shouldn't necessarily change any of the dialogue. Low-budget productions usually have better stories than special-effects-laden epics because more time and thought has gone into that arena. If you have a great story, tell it and look for areas to save money in the production.

Love stories or romantic comedies aren't usually remembered for their huge budgets (unless the actors want trillions); they're remembered for their story. My wife would rather see a love story over a Jackie Chan or James Bond movie. She tells me it's because of the story, and to a point that's true. Both Jackie's and Jimmy's films are stronger in the special effects than story. But any low-budget video you've seen will be quickly forgotten if the story stinks. Whether you shoot commercials, corporate videos, documentaries, or anything else, the story or script is number one. "Words are the glue that holds a low-budget video together." Wow, my first quote!

Of course, bad actors can murder a great script, but words still are your best tool to make the viewer forget you have no money. My first great science fiction epic, *Forever Was Yesterday* shot for $600 (film processing cost) back in 1978, was carried only by the acting. Not being able to afford sync sound for spoken dialogue, the entire piece is narrated. In film school, this is called a cheap copout to convey information to the audience quickly (and should never be used). But it is a cost- and time-saving tool. All of the sets were cheesy and free, the costumes homemade, and the space ship was my reading lamp. I shot all my space scenes in the Men's room in the library because it was dark enough (with the lights out) and the wall long enough to support the star field backdrop. When someone walked in to use the restroom, production had to be halted. Actually, people were more concerned about us shooting them using the bathroom than the planets on the wall. The epitome of no-budget production can be seen in Figure 2-3, and try not to laugh at the starship in Figure 2-4.

If you're resourceful, you will find ways to cut costs that are believable. Instead of telling the story from beginning to end, break it up. Start in the middle, flashback to the beginning, and jump to the end, all without confusing your audience. Try a nonlinear approach (not telling the story from A to Z). Writing instructors have told me "start with E, skip ahead to Y, then go back to B, and so on." This will confuse some, because 90 percent of the scripts you read are linear. If doing a corporate video or commercial, you are stuck with linear. But almost any other video may be more disjointed (not out of whack).

Figure 2-3
Forever Was
Yesterday

Figure 2-4
A state of the art
lampshade

One way to save money is in the location. A well-known landmark usually demands some kind of remuneration when approached if only to handle crowd control. An upcoming section will discuss how to find the free locations because that will help your budget the most.

It's All in the Eyes

Along those lines, let your underpaid actors do the work. Allow the script to furnish the background information without having to visually show it. If the script is written well enough, the actor may talk about a place without the audience having to see it. If you shoot the scene in close-up and the actor's expressions work, her visit to Mount Rushmore last summer may be shown through her facial expression without having to drive there and shoot. "I fell in love at Mount Rushmore," she said while gazing deeply into space. As the camera zooms into her eyes, we listen to her speak about her love. We've all heard about Mount Rushmore and know what it looks like; don't drag your actors there to shoot the scene. If you need dialogue between the lovers, shoot it in a Mount Rushmore-like setting (any park will do). The same thing applies with any landmark. Without traveling to a location or erecting a matte painting, the locale may be staged elsewhere. Hollywood does this constantly, and if it's okay for them . . . . When we landed on the Moon in July 1969, a low-budget producer would have had the astronauts talking about it seated in chairs rather than showing it.

Is It Real or Plastic?

It's best to avoid the phony sets if you can't make them look real. Plastic or paper rocks painted do not look like the real thing. Your mission is to make the inexpensive look like the genuine article. Today's viewer can spot low budget a mile away. Disguising that from them while still saving money is your quest. Look at the planet sets on the original *Star Trek* compared to the later versions. The original sets are laughable at best because we have become more sophisticated and can spot a 1960's *Batman*-type set a mile away.

The new *Star Wars* movies look far superior to the 1970s and 1980s versions. The older ones looked great and sported a large budget, but technology has made improvements that now make great look not as good.

If the audience is left to imagine something, their minds will create it better than Hollywood ever could. Look what happens when you read a book (not this one, a novel). You clearly imagine billion-dollar sets, name superstars, and special effects unmatched by anyone. Your humble scripted video can accomplish the same thing if done correctly. This may be difficult to pull off effectively, but great acting and believability are your tools.

An Imagination Is a Wonderful Thing Not to Waste

To save money, I've written and shot horror videos where no one ever sees the "monster." If done correctly, it works. The audience will feel cheated if you never show anything, but just glimpses, bits, and pieces of the creature work. I never had to create an entire monster. Look at successful Hollywood horror where you rarely see what's scaring you. It's more frightening not to see it because your mind has made it far worse than it often is. Nicole Kidman's film, *The Others*, took this approach. The acting was superb, the writing well done, and we never really saw anything. Imagine that, a horror film without the high-end special effects, gore, or floating ghosts; you can do it too. Once again, if you let the viewer imagine more, your cost will be less.

If your low-budget production is horror, look at the moneymaking Hollywood projects done on a shoe string and observe what made them work. The original *Halloween* was produced for $150,000 and made millions all with a no-name actress, Jamie Lee Curtis, the first great use of a Steadicam, existing sets and locations, and editing that made you jump.

Editing is another tool in no-budget production that saves money and can make people see what they never did. In film school, we studied Alfred

Hitchcock's *Psycho* by watching it a billion times. Everyone knows about the shower sequence and how Janet Leigh gets attacked in the shower. Through extremely fast cutting, it looks as if the knife penetrates Janet Leigh's stomach—you lose a lot of great actors that way. The knife stopped just before contact and the editor cut to blood flowing toward the drain. If you show just enough, the viewer will follow through with their imagination. Let the viewer work a little; they really want to but will never admit it.

In the same film a little farther along, Detective Arbogast is looking for Mrs. Bates and climbs the stairs slowly. She jumps out and stabs him. We never see the stab (which can be created in close-up by sticking a knife into a sack dressed in clothes filled with stage blood). In order to lower the gore (this was 1960), we see a one-second shot of the knife moving from above her head to thrusting down, and the editor immediately cuts to a thin splash of blood across Arbogast's face. We do the math and know he was stabbed.

Instead of hiring a stunt man (more money) to fall backwards down the steps after being stabbed, we see Arbogast moving slowly backwards, about to lose his balance with his arms flailing and the background moving— disorienting but effective.

Using People

We all know people in interesting jobs and places. Why not use them or their access to locations? Don't take advantage of them, but ask if they may be willing to help you out. All the contacts you've complied may be used. As an example, a client once wanted a video showcasing the earth's destruction in the year 2000. Everyone believed the "Millennium Bug" was going to cause computers to fail on January 1 and send the world into chaos. The dimwit who hired me wanted to show the fire and brimstone destruction of the planet and have an alien land and greet the audience (a great way to open your sales meeting). He had no money, the budget was squat, and where was I going to find flying saucer footage? As with any no-budget production, I broke everything down into smaller steps. First, the destruction —I purchased stock footage from Hollywood films of the end of the world (*Independence Day, Deep Impact*, and *Asteroid*). By cutting together bits and pieces, I had New York and Los Angeles wiped out for a million dollars less than they spent.

The flying saucer was slightly harder to locate. I searched every 1950s American International film ever made, but all that footage was black and white. I found a low-budget Canadian movie that was perfect. The saucer lands on earth and the door opens.

Figure 2-5
Alien meeting

Figure 2-5
Alien meeting

The last obstacle was the alien talking to the audience. Remember, smoke will save your butt if you have nowhere else to go. After the footage of the door opening ended, the house lights dropped, fog filled the room, and the alien walked out. Figure 2-5 shows your favorite alien.

The room was foggy, the lights very dim, and the papier-mâché head never looked real. The alien's voice was on tape, recorded by sticking a microphone in a metal trash can and flanging slightly in postproduction. For very little money, the audience was entertained. They didn't see a heck of a lot because of the fog and lighting, and the world didn't end on 1/1/01.

I feel compelled to give you a little more detail on the stock footage. Getting great material that someone shot may be expensive. Shop around and find the best deal. If you aren't broadcasting the final result, the prices are surprisingly lower. Tell the stock footage house your budget. If they can make anything work, they will try ($1 is better than none). Once again, don't borrow, steal, or use this footage in any other way than agreed upon. That cell is still waiting for you.

Never Mind, I'll Do It Myself

That last great effort in saving money in low-budget productions is doing it yourself. You obviously can't do everything alone (don't even go there), but if you at least take a role in all aspects of production, money will be saved. This may seem like common sense but may also lead to overproducing.

I worked with a client who overproduced and micro-managed to the point that his productions always cost three times his budget. In this case, if he let the professionals do what they were paid to, he would have spent far less.

If you are involved in all facets and watching the costs, the end total could be less. Keep a running total of everything spent and why; you need to know where the money is going. Other tools are deferring salaries, borrowing equipment for a credit in or copy of the final program, and selling shares in the production. If you write the script with the cost of every shot, scene, and location in mind, your bottom line will be lower.

Rent as many low-budget videos as you can to watch what works and doesn't. How could you have done that better? Anytime you are restricted to a low budget or none at all, you need another avenue to come up with a solution.

The most important thing I can tell you is don't get discouraged when your best ideas get rejected by the client, and they will. What do they know anyway? Keep them apprised of each cost as it happens so your no-budget doesn't become a monster. If they see an expensive shot coming, maybe they'll nix it. In the end, it's all up to a higher power . . . the client!

Are Storyboards Really Necessary?

If I had to sum up video production in one word, it would have to be *preparation*. Always expect the unexpected. Don't let them see you sweat. Remember the Alamo! These are all great words of advice, although slightly clichéd. The best way to prepare is to totally have your act together, and the best way to do that is at the beginning.

To be honest, I hate doing storyboards, my students hate doing storyboards, and few others enjoy them. Unfortunately, they are necessary. Students complain that they'll never use them in the real world. I work in the real world and I use storyboards all the time. I don't like them, but I need them.

After reading the "How to Get Ideas" section, you're now full of it (ideas). So now that you have an idea for a video, it should be storyboarded.

All feature films begin with a concept, a script, and then a storyboard. With all that goes on and into a feature film, it would be impossible to plan everything out in detail without a storyboard. A video production of any size requires no less. Just because the running time of a video may be less than 90 minutes, you are still telling a story that needs to be planned out in detail.

I'm assuming you already know what a storyboard is. My purpose is to help you create a better one for video production. First, I'll discuss why I think a storyboard is necessary in any production. Most low-budget videos don't have the elaborate special effects that feature films possess. Just because a car isn't blowing up doesn't mean that you don't have to plan where each person is at a given time.

The length of the production makes no difference. On a sheet of paper, place the visual instructions on the left and the audio information on the right (if you live below the equator, you may switch this order around). In film school, I was always told to "label everything, everywhere, always."[1] If you don't, questions and misunderstandings will arise that will bite you later. On the top of the form, list the pertinent information such as the client's name or account, the working title of the video, the duration, date, page number, and any other information necessary to set this particular work apart from the hundreds of other projects you may have.

Some people believe you have to be an artist to create a storyboard. It does look a lot neater and more professional if you have accurate, four-color, lifelike drawings depicting the action. But unless you have to impress the client with that type of detail or you've always wanted to enter those contests in comic books or the back of matchbooks, that type of detail isn't necessary. What is important is that all the video information is recorded in the video column. Whether it is drawn, typed, printed, computer generated, or in Braille, whatever works for you and the client . . . do it.

Because I can't draw a straight line without legal help, I type out my storyboards. If I'm feeling creative, I will cut out pictures and attach them to the storyboard. In the case of a current project, I used a little of both. I cut out a picture of a car and wrote above it what I wanted to happen visually.

---

[1]Richard Curilla, The Pennsylvania State University, 1979.

All camera movements (zoom, pan, tilt), actor movements (female walks up and gets into car), and prop and set information (one 1922 Model T Ford, a paper bag of groceries, and a crowbar) are recorded in this column. I prefer to write too much detail in this column than not enough. If for some reason I wasn't at the shoot (of course, that would never happen), someone else could look at my storyboard and know exactly what I wanted. If you spell everything out, no one will misunderstand or question it later. Sometimes things look great in a storyboard, but at the shoot it's another story. There's no excuse for poor planning. Come to the shoot prepared with your storyboard, and if you come up with something better on the spot (no pun intended) at least you have something to use as a guideline.

I grade my students on the amount of information in their storyboards. They receive higher grades if they include too much rather than too little. When clients give me hand-drawn, stick-figured, ambiguous storyboards, I wish my students had to work on this project so it would be done correctly.

As a director, I'm not the only person who sees the storyboard. Camera people, editors, continuity people, prop people, set dressers, grips, narrators, actors, and many others all rely on this piece of paper. Other than the script, it has the most value in a production. Even when the shooting is completed, the graphics people and the editor will also need that information. If you do change something at the shoot, get into the habit of making the change onto the storyboard. The storyboard should be accurate and up-to-date all the way through the production/postproduction process.

The other side of the storyboard contains all the audio information. Type all spoken words, whether it's said on or off camera, in upper case; it's much easier to read. In addition, spell out all the words. Don't use abbreviations. Some literal actors will read the abbreviations and pronounce them as such. It's also important to give as many timing cues as possible. If a narrator has to say three sentences in eight seconds, spell that out in the storyboard. If you cover your butt at every possible opportunity, you'll never get sunburned down there.

If an actor is supposed to read something a certain way, say that in the storyboard. This also gives the talent some guidelines.

As always, your first draft of the storyboard will probably look nothing like the final copy. But by looking at the storyboard, you can see how the video has evolved from just an idea in the back of your mind to a completed production that is making the client rich (unless his name happens to be Dave).

Review: The 10 Most Important Things to Remember When Storyboarding

The 10 storyboarding points to remember are as follows:

1. Have separate columns for the video and audio. Don't overlap or confuse the two (they don't like that and will fight).
2. Label everything, everywhere, always. If you list everything on the storyboard, you'll live a longer, happier life.
3. Be as detailed as possible in the video column. List or draw actor and camera placement, movement, set description, and props.
4. Create the scenes in chronological order. It has to flow like the finished video. This sounds like common sense but follow it.
5. Put everything relating to audio in the audio column. Everyone on the crew who has anything to do with sound should be able to look in that column for the answers. (The same applies to the video column.)
6. Use upper case (capital letters) for all spoken audio. Use lower case or italics for other information.
7. Spell every word out. Don't abbrev. (Isn't that clever?)
8. List timings for every piece of audio. If something needs to be said or read in eight seconds, it's the talent's responsibility to make it happen in that amount of time (with a little help in editing).
9. Let everyone involved in your cast and crew see the storyboard and understand it. It shouldn't be a secret document for the director's eyes only.
10. Although your storyboard is a blueprint, it's also a guideline. If you find something that works better at the shoot, do it. But you still need the storyboard as a guideline to follow.

Now that you know everything there is to know about storyboarding, it's time to learn some really exciting techniques!

How to Find the Right Location: Free Ones

Most of your video budgets are relatively tight, so if you can save a few dollars in one area, do it. One of the ways in which most people try to save money is by shooting on location. If the location if far from your business, then the costs of travel, lodging, and outside help can make your costs higher. But if your shooting location is close by, money can be saved by shooting there rather than using a sound stage or building a set.

When a client comes to you with a project, one of the first things to decide is the most cost-effective place to shoot the video. You really only have three choices: shoot at the client's place (factory, shop, or restaurant), your place (your studio), or on location (someplace else). Any area that isn't your place is considered a location.

So how do you find the right location? Is it better just to shoot everything in the controlled environment of a studio, or do you want to deal with the problems that location shooting might bring? Hopefully, we'll weigh both of these decisions carefully and you can decide which is best for you.

You know what's involved in shooting at your place of business. You know your assets, your equipment, and your strengths and weaknesses. Shooting there doesn't really host a lot of problems. You know it well and have used it often. Maybe you just want to try someplace different or have an idea of where you'd like to shoot. A location can be some exotic port of call, or it can be five feet away from your front door. Now that we know exactly what a location is, how do you find the right one?

Cost usually is a determining factor. Will the budget allow you to travel with equipment, talent, and crew? How far can you travel for the $43 you've been allowed?

Life's a Beach

Recently, I shot a furniture commercial on the beach. In order to get the concept across to the viewer, the sofa had to be on the sand at the seashore (see Figures 2-6 and 2-7). With digital effects, we could have shot the furniture in the studio and supered it over a shot of the beach. But unless that's done correctly (usually requiring a larger budget than we had), it looks phony. The shot had to be done at the beach.

I talked to various chambers of commerce, tourist bureaus, and other agencies that knew the best locations in the area. These services are free and can be very helpful. By looking at brochures of various locations, you can weed out the places that definitely will not work.

Most furniture commercials utilize the usual slow dolly shot past a couch, dining room set, or the like. These spots are usually shot on location in the furniture store or in the studio. Also, attractive models are usually lounging on the particular item, caressing its fabric, or a family is shown

Figure 2-6
Furniture on the beach

Figure 2-7
Furniture on the beach with more people

with their two-point-five kids spilling popcorn all over the sleeper sofa while watching videos on their wide-screen TV. But aside from that, everything usually remains exactly the same.

I approached our client and asked if I could do something different with his next commercial. Why not take the particular piece of furniture out of the store or studio and shoot it in some kind of exotic local. I had originally envisioned having the furniture against the backdrop of a beach in the Bahamas, Tahiti, or some other equally exciting port of call, but our budget for local television precluded that option.

After convincing the client that we would produce a new and exciting 30-second furniture commercial on time and within budget, I finally had his attention. My crew and I requested that he allow us to use one particular piece of furniture for a day, and let us make a spot that would stand out from the pack.

Let the Fiasco Begin

The best way to find the right location is to actually look for one. In my case, this involved going to the beach. I must have looked at 50 beaches before I found the right one for the shot. You know what you're looking for, and the only way to find it is to actually begin the search. You can hire people to look for you, but half the fun of location scouting is just that: scouting. Take your compass out of mothballs and go.

Something to look for when scouting is how accessible is the location to the cast, crew, and equipment? It's no fun to lug tons of equipment and people across a frozen tundra, tripping over penguins (that's another story I'll discuss later) only to find it's now too dark to shoot. Can you get permission to shoot at this location? If you get permission, will it cost anything to shoot there? Some producers with very small budgets (sounds like an affliction), try to shoot someplace without telling anyone. Sometimes this works, but I wouldn't want to chance it. Far too much is at stake to risk it all for a few dollars or someone's OK. Although you won't need permits on every location, someone has to be aware that you're shooting there. Offer to get a permit. If they say you don't need one, make sure you have written authority that you can, in fact, shoot there.

Another thing to look for on location is accessibility to power. I don't mean is there a general, king, or president around. Can you run AC lines, do you need a generator, and so on? Decide what you'll need when scouting, not when you're at the shoot.

Preproduction takes more time and can be more important than the actual production. Work out all your possible problems and have backup plans ready. I've been promised the world (I don't really know what I'd do with it) on a location scout/preplanning trip. When we arrived for the shoot, some of these promised things didn't happen. That's where contingency plans come in. If you have to have that elephant drop the Ping-Pong ball on the aircraft carrier's deck, make sure you have access to another elephant if Jumbo is sick (and another Ping-Pong ball if Jumbo steps on the first one). By the way, you wouldn't think that an animal that eats nothing but straw and nuts would have a gas problem. That, too, is another story.

My scenario was simple. The scene would open with a young business woman having a terrible day at the office. Everything is going wrong. She spills coffee on her blouse, misses all her deadlines, gets chewed out by her boss, her proposal gets eaten by the copying machine, and she leaves work in the pouring rain only to find that her car has a flat tire. After these and other things happen to make her day really miserable, she returns to the sanctity of her home to relax on her favorite sofa in her own living room. As she plops onto her prized sofa, the camera pulls back to reveal that the sofa (and the living room) are on the beach. As the store's logo appears over the ocean, waves pound against the shore and a seagull cries out as it flies by.

In order to make all of this happen, the videographer and I spent a day traveling to the closest beaches. We tried beaches that surround natural and man-made lakes, but the sand and the waves just didn't have the same feel we were looking for. The only place to find real waves and a real beach is at the ocean. Since we reside in landocked Pennsylvania, our nearest beaches are in New Jersey, New York, Maryland, and Delaware. Because we were shooting the spot in late September, we believed we would have little trouble finding a secluded beach.

We were finally able to secure the perfect beach in Ocean City, Maryland. The other beaches we had seen were either too crowded, had debris scattered along their perimeter, or had too many logistical shooting problems.

On the day of the big shoot, we loaded two vans, one contained the crew and camera equipment, and the other housed our living room furniture. Three hours later, we were parked at the edge of the white sand, ready to carry all the furniture piece by piece down to the water's edge.

The split second our two vans arrived at the beach, the local police came screeching up behind us in their squad cars. Even though both of our vans had the station's call letters spelled clearly out over every flat surface, the police insisted on asking us what we were doing at the beach and where we were from. It was only after I explained to the officer that we had the necessary permits and gave him our contact person's direct cell phone number that he began to believe we were really here to do what we said we were going to do.

After our brief brush with the law, we were ready to begin. As we proceeded to assemble our living room, we began by stretching a 10-foot piece of five mil plastic down on the ground to protect the fabric from the sand and ocean spray. As our audience of beachcombers and transient tourists watched us decorate our beach-front home, we were asked just exactly what we were doing. It seemed as if no one had ever seen anyone carrying furniture down to the water's edge before.

By the time our set was completed on the sand, we had a realistic, functioning living room. The power for our lighting equipment was run from a DC-to-AC inverter in our production van. With enough extension cords we were able to power our lights and even illuminate the practical lamps on the end tables.

While setting up our living room, we had also noticed that several vacationers were watching us through their binoculars from the high-rise hotels that dotted the edge of the beach. I have to admit that seeing a couch, love seat, coffee table, end table, and lamps sitting out in the middle of the beach with the ocean as the backdrop looked rather amusing. At least the tourists would have something to tell their loved ones.

Even in late September in Maryland, the F stop on the beach is between F16 and F22. In order to cut down on the sun's incredible brilliance, we erected a 12-foot by 12-foot silk shade 25 feet in the air above our sofa. With the strong breezes blowing off of the ocean, our silk immediately became a huge sail. One of the crew instantly scrambled to grab a sand bag from the van but remembered our location and instead decided to act as a human sandbag. While standing on the legs of our beefed-up C-stand, he was able to steady the base of the stand as the silk itself rippled violently in the 30 mph-plus winds. With sufficient shade on the sofa, we then had to contend with the four stop difference between the sofa and the white sand. The sand immediately behind the couch was glowing white. Without having access to lange lighting instruments like 12K HMIs or Maxi Brutes, we had to resort to using every light in our arsenal to try to match the background with the sofa.

Our female model hadn't seen the sun the entire summer and was as white as our overexposed beach. By gelling our tungsten lights with straw to medium amber gels, we were able to bring up her flesh tone slightly higher to that of a living appearance. If we had used dichromatic filters on our lights to balance the color temperature correctly to that of daylight, we would have lost valuable wattage as well as having a blue ghost as a model. The slight warmth that the gels provided (and the uncorrected color temperature from the tungsten lights) really added to the scene by allowing the model to stand out from the sofa. We also used a gold reflector as well as several sheets of foam core to bounce the harsh sunlight onto our subject. Because we still wanted the initial shot of our actress falling onto the sofa to appear as if it were happening in her living room, we had to shoot everything in a medium close-up to avoid giving away our humorous ending.

As the model flopped down upon the sofa in exhaustion from her bad day, the camera would dolly back to reveal our actual location. As soon as we began to shoot, the PVC dolly tracks would immediately become covered with blowing sand and make the movement very bumpy. I never thought that a grain of sand would be felt under the weight of a fully loaded dolly. To solve the problem, we had one crew member wiping sand from the dolly track with a towel, the client raking the sand under the model's feet between takes, and a third still clinging to the C-stand to keep the model from being crushed.

In order to get the model's movement exactly in sync with our camera's pullback, we had to shoot the scene many times. I would be yelling out cues to the model to "fall," to the camera to "pull back," and the seagulls to "speak." Tourists would somehow sneak through the crowd surrounding us when we were shooting. This would always spoil the shot. I guess near-sightedness is very prevalent with people who frequent the beach. They couldn't possibly see what was going on 15 feet away; they had to practically sit on the sofa with the model to understand what we were doing.

Other great ways to spoil the shot included having a seagull attack the camera operator at an inopportune moment thinking he was food, someone looking directly at the camera and ask if we were making a commercial on the beach, someone comparing their Sony camcorder with our Betacam because they were both made by Sony, or something else would cause the shot to be less than perfect.

People who would normally walk down along the ocean in their bare feet with extended stomachs and circa 1950 swimsuits wouldn't notice the set lights until they were in the middle of the shot. It would only be at that time that they would freeze in their tracks like deer in headlights, scratch themselves, and slowly back out of the shot. It all looked very natural. In hindsight, I now know that I should have placed our teenage crowd control people farther down the beach to stop these stargazers from ruining takes.

As our model plopped down onto her favorite couch over and over again and thought of her happy place, my crew was beginning to complain about the sun- and windburn they were receiving. It's hard to believe just how fast time flies when you're working on the beach. Just walking back and forth along the sand (let along carrying furniture) was like walking with 10-pound weights around your ankles.

In the end, we added the other audio sound effects, voice-over narration, and music. People who weren't as close to the project as me were asked to screen the spot and offer their comments and suggestions. When the client saw the completed spot, he knew that he had something that was different (that means he liked it).

I often ask myself if all the aggravation, time, and no money spent were worth it for just five seconds of some living room furniture on the beach. Each time I answer with a resounding . . . yes!

You may not believe it, but as a video maker you have a lot going for you. People will actually offer their homes, business, or possessions free of charge if you simply point a video camera at it.

Be prepared for anything. What you don't prepare for, that's what will happen.

Review: The 12 Most Important Things to Remember When Looking for the Best Location

The points to remember when location hunting are as follows (you get two extra for free):

1. It may seem like common sense, but actually look at the location. Brochures are helpful, but really seeing the place can make a lot of difference.

2. Talk to tourist bureaus to get ideas of where you might shoot. Look at their brochures.

3. Take all the time you can afford and go to each location to see what it offers. Walk around and get the feel of the place. It's very important to really look around. Will that building that's blocking the sun be torn down tomorrow?

4. Once you find the best location, go there during the time of day you plan to shoot. Locations can look very different in the morning and the afternoon.

5. Make sure the location is accessible to cast, crew, and your 80 pounds of equipment. Pack mules cost extra.

6. Listen to the location. Will outside sounds affect your production?

7. Make sure you have sufficient power to operate your equipment. Find out what the location offers, and then bring additional power if needed. A spare general never hurts.

8. Get permission to shoot on the location. Obtain permits if required, but make sure someone important knows you are there.

(continued)

(continued)

9. If you'll be shooting at a public location, make sure the local police are aware of that. If they arrive and don't know why you're there, you may have another opportunity to share a cell with someone named Spider.

10. Have backups of everything, like another possible location if one floods or burns up. Because you're away from home, make sure you have backups of everything you may need.

11. Leave the location exactly as you found it, or better. Bad PR on location goes a long way.

12. Make sure someone from your home office knows where you are. If you need to be reached, it's better to be at the place you're supposed to be.

Just remember to check, double-check, and recheck the location as many times as necessary. Don't be afraid to ask for the assistance of others. You may be out of your natural element, but you can make that work for you.

Shooting with Nonprofessional Actors (The Client's Relatives)

One of my favorite parts of producing or directing videos is being able to cast the talent. Like Cecil B. DeMille, you look at thousands of faces and biographies and pick the ones that interest you. When you work with professionals, you can find the exact type of character you're looking for. That's what professional actors do—they act. However, sometimes you aren't offered the luxury of being able to hire outside talent. A case in point is when the client wants to use his family or friends in the video to save money.

I could go into a long song and dance on how the client shouldn't use nonprofessionals (amateurs) as on-camera talent when everyone behind the scenes is a professional. I'm not talking union here; it's just if you want professional results, you should hire professionals. The on-camera people are what the viewer sees; they never see all the work that goes on behind the camera. If Uncle Ralph does a poor job of being a convincing salesperson in the video, then the public is going to have a hard time believing him as an actor. The client then comes back to you and says the video didn't work and was a waste of time. If they had let us do our job in the first place, the program probably would have been well received. But the client is footing the bill and it is his/her/its prerogative to make that kind of money decision.

This isn't about griping to the client to hire professionals (not the guys with the bent noses). Try to talk them into hiring real talent. It can't hurt and maybe you'll win. But if you are stuck with relatives or friends of the client, this should help you get the most out of their performance (here I use that word very loosely).

Let's assume that the script has been approved and you're ready to shoot the project. The budget is so low (how low is it?) that the client has called in all his relatives and friends. All he has asked them is "Do you want to be on TV?" This is a question that few people refuse.

I'll try to take this one step at a time and cover some of the possibilities that you might run into in a situation like this. The first step is to get everyone together and explain what you're trying to achieve in the video. I like to role-play all the parts to everyone so they can see what I'd like them to do. You should have a clear vision of what each character must do, say, and act like. It's almost impossible to get a spectacular performance out of someone if you really don't know what you want. Once you've done your routine in front of the client's cast of misfits, don't ask if anyone has questions at this time. Everyone will have their own expert opinion on how they could perform the part better and add more life to the video. Instead, I prefer to get each actor aside individually and go over his or her part one on one with them. I do it as a group effort first so everyone can see how each part fits together as a whole, but the one on one is where the little nuances come to life and where I ask if there are any questions.

Once I asked if there were any questions in a group of relatives. One person asked a dumb question ("How much am I getting paid for this?"), which then made everyone else ask 50 questions.

With this one-on-one approach, I act out what I see the character doing, and then I ask the actor to try his or her hand at it. I find it better if I perform the role first. The actor usually has an easier time mimicking my performance than doing his or her own (often wrong) interpretation. A role can be performed in a million ways, but if you have something specific in mind, save yourself a lot of time and show it to them. In Figure 2-8, the client's sister gets ready for her big break.

If any of the characters have speaking parts, that involves a little more skill and patience on your part. Actors who are used as extras aren't much of a problem, but if they have to speak, they require more instruction. Have them say their lines to you in a conversational tone, just as if they were talking to you. Some will try to project their voices and talk very loudly; others will overact as if they are on stage in front of thousands. I tell them that

Figure 2-8
The client's sister

TV is a very intimate medium. It's very conversational. If two people are sitting at a table talking quietly, that's just how they should do that. The only difference is that a camera will be close by. The three most important things for nonprofessional actors to remember are to be yourself, be yourself, be yourself (or relax, relax, relax). Or go away, get lost, and good riddance.

Now that they have mastered the dialogue, don't try to have them do it all at once. If you break it down into smaller bits, they'll have a better chance of doing it right and feeling more comfortable with it.

Once every character is familiar with what they're supposed to do, shoot everything in sequence. It's very difficult for a nonactor to shoot something from the middle first, and then do the end, and then the beginning. The video may not be very long and you'll get more out of their performances if you shoot chronologically.

As a director, the friends and relatives the client hired need to know you're in charge. They can't say "cut" in the middle of a take because they don't like it. There will be laughing jags and nervous sessions when they just have to get it out of their system before you get anything done. Don't treat them like children either (unless they are children) and try not to talk down to them. They are excited about doing this and they will remember the experience long after you've forgotten it. Try to make it an experience they will enjoy. Let them have fun with the role and you will get much more out of them.

Always rehearse several times before you roll any tape. The red light on the camera, the quietness of the set, and them being "on" for the first time can make them uncomfortable. After you get a realistic performance out of them, roll the tape without telling them that the camera is on. If it's a great performance, I'll tell them I want to do it "for real" this time. If it doesn't work out because they know the camera is on, at least I got a good take earlier. If I get even a better performance out of them, I have a safety.

Always expect the unexpected and don't blow up at someone for not getting it right after 60 takes. They're very nervous, they want to please you, but some people don't have it in them. Instead, take them aside and try altering the script slightly. If you embarrass them in front of everyone, you've lost them as an actor. If this happens, make sure you check your brakes before you go home.

Review: The 10 Most Important Things to Remember When Working with Nonprofessionals

The reminders here are as follows:

1. Talk to them as a group. Explain everything about every character.
2. Talk with each actor one on one. Tell them what you expect and then let them do it.
3. Block everything out. Everyone should know where to be and what to do.
4. Rehearse everything, no matter how insignificant the part.
5. Tape the later rehearsals. Don't tell them the camera is rolling. Turn off the tally light.
6. Break each larger part down into smaller bits. This makes it easier to chew for the actors (unless they have big mouths).
7. Shoot in sequence. We don't live our lives out of sequence. Why should the nonprofessionals have to?
8. Be willing to rework difficult lines, movements, and positions. Sometimes a slight variation can solve the problem.

(continued)

9. Be easy on the actors. Don't overwork them, embarrass them, or expect miracles (unless they look like Moses).
10. Have fun with the experience. They certainly will.

I've been in this relative/friend situation more times than I care to remember. Just take a deep breath, keep calm (if they see you're calm, they are more apt to be calm), go slowly, rehearse everything, and have fun with it. Most friends/relatives think they are working with a top Hollywood director and will treat you accordingly. You never have to tell them the truth. I certainly won't.

Do I Buy, Rent, or Steal (Borrow)?

Now that you're ready to begin the low-budget video-making process, how do you get your hands on equipment? Having your own video camera does make the process simpler, but if you need better tools than what you currently own, where do you go?

Because video technology changes overnight, you should have the best equipment to enable you to work professionally. Does this mean you should buy that new three-chip camera you had your eye on? This is a tough question to answer and involves contemplation. Owning a new camera means it's yours and you may use it and charge for its use. But in low-budget productions, who is going to foot the bill to pay for it? The same thing is true with lighting equipment. You probably won't purchase 20 HMI lights if you have one shoot in a stadium. But if you used HMIs often, are they worth the purchase?

Cameras Are the Eyes of the Soul

If you invest in a lower-end, three-chip camera (I'll only be talking about professional equipment here, and one-chip cameras are inexpensive), you are shelling out a minimum of $2,000 currently. With that same amount of money, you could produce several TV spots and a short corporate video (with my grand, low-budget approach). My advice is don't buy new equipment, although tempting, until your work load justifies it.

Recently, I was asked to shoot a wedding for a friend I had known for 15 years. She wanted an extremely high-end wedding using a digital three-chip camera, intensive editing, and the final product on DVD. I personally own two analog, one-chip cameras and an *nonlinear editing* (NLE) system. Everything else I would have to acquire for the shoot. Her parents were footing the bill and cost was no object (don't you love rare people like that?). I couldn't very well purchase a new camera and DVD burner and charge it to them.

I had no other weddings on the horizon so buying a camera and DVD-burning system wasn't in the cards. I contacted another video friend (don't you hate people like me?) and asked to borrow his three-chip camera. After all, I am a professional and will treat the equipment with respect like it was my own, and he has borrowed my audio and lighting equipment in the past.

In situations like mine, the first step instead of purchasing is to see if you can temporarily acquire the desired object or borrow it to keep the costs down. If you damage or lose anything, you should be responsible and pay for it, but borrowing will save the budget and allow you to use the equipment you need.

The other option is paying to use it, commonly called renting. With video equipment, this involves determining how long you want the item, paying a deposit in case you skip town, and being insured adequately. This makes the budget higher so this is a secondary option: borrow first, rent later.

When you borrow anything for a video production, you ought to do some common-sense things in return. Since no money is exchanging hands you should return the favor if possible, purchase all the needed peripherals such as tape and expendable batteries, and return the equipment in better condition than you received it (fully charged batteries, clean lenses, and an organized camera case).

Although I mentioned borrowing a camera, the same principles apply to anything borrowed. The most critical item in borrowing something is knowing when to return it. Since you have something of value from someone else in your possession, ask beforehand when they would like it returned. If possible, bring it back early if you are through using it. The borrowee may need the equipment when he or she wants it returned, and being late could jeopardize their other work.

Another common-sense thing is not to loan someone else's equipment you have borrowed to another individual (subletting). I won't go into the problems this creates. Don't show some six-year-old kid "your" new camera and let him play with it. The "you break it, you bought it" rule applies here too. Basically, just do what you would do if it were your camera, not a friend's.

Any no-budget video program can benefit from the borrow or barter system. Determine what your desired piece of equipment costs to rent and you will see how much you have saved on the bottom line.

Luckily, I have an editing system so I didn't have to borrow that for the wedding. The DVD-burning system was still at the expensive stage at the time of the wedding, so paying someone for the use was much cheaper. This wedding involved me doing all three: buying—I purchased an editing system for another job and used it for the wedding, renting—I paid for the use of the DVD burner, and borrowing—I bartered the use of the camera and my friend used some of my stuff. The only thing I did not resort to in this arena was stealing.

I Want My Own Toys

Let's look at some of the pros and cons of owning equipment. Beginning with the pros: You now have the item to use anytime your greedy heart desires, new equipment functions well and should be dependable, and you don't have to worry about getting the rented or borrowed equipment back at a certain time. If the project goes longer than expected, it's your stuff. Lastly, you can market yourself as having your own equipment. You shouldn't tell people you have your own equipment if you don't. If a sudden project arises and you can't get your equipment, that makes you look unprofessional. As I mentioned, I don't own a high-end digital camera, but I do tell clients I have access to one.

By wording it as "having access to" means you should be able to get the equipment on short notice. When I see "have own equipment," I always picture it as being well worn, beaten up, and ready to be replaced. Actually, that is rarely the case. What it really means is the person's charges may be slightly higher because they have to pay off that new Model XXX 4A.

That brings us to the negative aspects or cons of owning equipment. When you own something, everyone in the business is going to want to borrow it. This depends on how you feel about it, but renting may help pay for the equipment. I just turned a negative into a positive!

My boss personally purchased a new three-chip digital camera to replace the company's light-sucking, 1980s vintage broadcast camera. The old camera worked perfectly, but the tiny digital unit produced better images. By renting the camera (and me) out to the company, he paid for his camera in rentals after only a few shoots. He now owns the new camera and still makes income on it, and it didn't cost him anything.

Another bad thing about owning is, unlike my boss, paying for the system: the overhead. Another friend of mine owns a high-end NLE system. Wanting and getting the best on the block, she invested a ton of money in the unit. Now she has to charge more than the usual rate to pay off her lease. Without having the work to justify that type of expenditure, she has to make more money just to break even.

She claims this to be the Catch-22. She needed the NLE system to get the job (she really didn't want to pay rent to anyone else), but she can't get work because she has to charge more per hour. Some rearranging of priorities might have helped in this situation.

Owned equipment will malfunction at some time and it's up to you to make it well again. Already instituting a maintenance schedule, you now have to get the equipment fixed and may lose money until that happens. That also brings to mind liability, theft, dropping it, or letting someone borrow it who shouldn't.

Owning any item that will receive wear and tear has other negative aspects as well. Cameras are subject to more abuse because they are mobile, but anything can be potentially damaged from use.

Rent-a-Thing

Renting is a more expensive way of borrowing an item. Paperwork must be filled out and signed; deposit money must be shelled out and then refunded. Availability, knowledge of its use, and insurance are all factors you must deal with when renting.

With the pros and cons again, the good thing is you get newer, well-maintained equipment without having the associated expense, you can afford to rent something you could never hope to purchase, and the liability is gone as soon as you return it.

I had a high-definition shoot where I couldn't afford to purchase a $100,000 camera, but the $1,500 rental eased my conscience. I was the only kid in town to use the camera and I could impress everyone that I knew how to say the name of the camera 10 times fast. With something very new or highly expensive, no one in their right mind will let you borrow it; renting is your only option.

That brings to mind the question, should I rent or lease to purchase? This is easily answered: If you will be using the object enough to pay for its costs over a set period of time, rent to own. I use HMI lights every fifth shoot. At several thousand dollars a unit, hundreds of dollars for a bulb, it doesn't make sense for me to buy or lease them. I pay about $100 to rent each light and return them when finished. I have no room to house them (Fido can sleep outside), a roughly handled light's bulb will blow, and by the time I paid them off in a lease to own, they'd look like war surplus.

If you do your homework, it shouldn't be difficult to figure out which option works best. Every piece of equipment, job, and situation are different. I can't tell you "in this case definitely buy." I could, but if I was wrong, you would hate me.

With responsibility comes price: Your bottom line rises when you rent, you might not be able to get what you need when you need it, and someone might have to teach you how to use it.

My best "borrowing versus renting versus purchasing" rule is if the video work will warrant the purchase or pay it off, you should purchase. If it's something that you may use a few times, unless the price is next to nothing, it doesn't make sense to buy it. The clear winner here is borrowing. You can now get all of the gain without any of the pain. The hard part is finding someone from who will let you borrow the equipment. If you happen to see me coming to your business, that may be what I'm after.

I've purchased large pieces of equipment for a specific job only to have that account suddenly end. I was stuck with equipment I couldn't afford to keep. What do you do when the unthinkable happens? That's why God gave us Ebay. The stuff is now spending the rest of its life living happily ever after.

The TV Commercial

One of the easiest things to determine with the untrained eye is the difference between a national TV ad and a local one. I know that the budgets are significantly higher in the national ads, but a low-budget local commercial doesn't have to look cheap (I should know; I've done enough of them). Hopefully, everyone will agree with me in wanting to make their spots look like good national commercials.

While working for an NBC affiliate, I've done over 600 local commercials for our station and I think most of them could compete with the look of a national spot. Obviously, the biggest differences between national and local spots are the budget, name actors, animation, special effects, graphics, a realistic time frame, state of the art equipment, pools of talented people, and the budget. I know I mentioned budget twice; that's really the biggest difference in most people's eyes.

So how does a local TV station create an opus that looks like it cost a million bucks? You really have to start at the beginning with the concept. Many national ideas are downright silly, but a lot of times the viewer will remember something goofy because it stands out from the rest of the pack. You may not have a team of high-end thinkers at your local station, but a simple concept can be bounced off other creative minds. That's how the big ad agencies do it.

It's extremely important to storyboard no matter how small your budget might be. Once it's written down on paper, you can expand or subtract from that concept. Run it by the client. Since they're going to be paying the tab, they have to approve the idea before you really get started producing it. Even if the storyboard is handwritten and drawn with stick figures, it's the map that everyone in production and postproduction will be following. Tip number one: Storyboard every spot.

You may ask how can a storyboard possibly affect the way a finished spot will look. If everyone isn't on the same page throughout the production process, or if corners are cut, the viewer will probably end up noticing it.

Production values are the main things that the end viewer will see. This is the one place where you shouldn't scrimp if at all possible. Of all the local commercials I produced at the station, only three were shot on film because of budget. Film obviously looks better than Betacam or digital video if used correctly. Joe Average at home isn't going to say, "Hey, Martha, that new Wilber's Market spot was shot in 35mm film!" It will look different, but he might not know the reason why. If the budget allows, try shooting the spot in film. If you can't afford it, use DVCAM.

Even high-end, analog Betacam can look like monkey vomit if it's not lit correctly. Many times a crew will just throw a 600-watt Omni up, blast it at the talent, and say the spot has been "lit." Yes, it has been illuminated, but it doesn't look good. Take the time to set up lighting. Every station on earth has more than one light; use them. Try three-point lighting, and use diffusion, silks, and gels. Make the actresses look modeled. Don't be afraid to use a hair light or eye light. If shadows are offensive, try bounce lighting. Never just pump up the gain of the camera to substitute for good lighting. It does make the scene brighter, but also a grain fest. So tip number two: Light all the shots.

A moving camera is another thing that separates the pros from the nonpros. A pan, tilt, dolly, crane, or jib shot easily puts your spot in a different category. Notice that I didn't mention a zoom. A slow creeping zoom can be effective, but the overused zoom is always too great of a temptation to most. Anything with wheels can be used as a dolly. I've used skateboards, wheelchairs, and shopping carts. The viewer appreciates moving camera angles. Tip number three: Use a moving camera whenever possible.

It's tough to find great on-camera talent for nothing or next to nothing. Most bipeds on this planet are trainable or directable. It just takes patience and time. The latter is really a luxury in TV production. Sometimes you're stuck with the client's relatives or family. Make the best of it and try to make it a fun experience for them. Check local theater groups. Most of these actors will pay you for the chance to be on TV. But you've got to rehearse to get the talent to display that "talent" on the screen. Tip number four: Rehearse the talent and try to make it enjoyable for them.

In editing, some will try to use every special effect their *Ampex Digital Optics* (ADO), Video Toaster, *Avid nonlinear editing system* (AVID), or whatever has. Use a little decorum in this area. Shots flying around all over the place can look pretty silly. Don't show these nifty effects to the client because they'll want to use them all. Most effects only call attention to the effect itself, not to the product, place, or person really being represented. If the effect actually makes the shot look better or if the shot needs it, try putting it in. How many shots of large fish eating smaller fish do you see in the national spots? Tip number five: Lay off the cheesy special effects; instead use them on the home video you show your kids.

Sound levels that are difficult to hear, are all over the place in level, or are just plain muddy screams "local!" Get a good level when you shoot or do narration. Mix the music under the voice. Try stereo if your station broadcasts in it. The viewer will notice the difference. In fact, most will notice good sound before they will a good picture. Tip number six: Sound should be heard, not seen.

Consistency in onscreen graphics also differentiates the small fries from the biggies. Our art director created all the graphics in all our spots to have a specific look. If 15 different people are creating spots, you'll have 15 different looks. Some will look better than others, but you should be consistent. In food spots, try to keep the same font style and size in all spots for the same client. If one person or a group approves or at least looks at *all* spots, this consistency will be maintained. I used to hate having my work being scrutinized by a group of my peers every week (even after the client loved and approved the spot), but it really makes a difference. A lot of little things I would miss would be caught by someone in the group. Some of my commercials have been award winners because of this group support and station spot consistency.

This can be taken to the extreme in that every spot looks like it was stamped from a cookie cutter. I don't mean that. Individuality is important, and every spot that your company produces should not look exactly the same. But within the spot, be consistent. Don't change fonts six times. I think you get the picture. Tip number seven: Be consistent (I think I've used that word consistently enough).

Try letterboxing your next spot. Once *high-definition TV* (HDTV) becomes the vogue, most commercials will be shot in the 16:9 aspect ratio where the vertical is nine units high by 16 units wide. Be ahead of the pack. Shoot the spot as you normally would, but use the wipe function in post-production to create the magic black bands. No one else in your market is shooting that way yet and it will stand out from the rest. MTV has gotten away with this technique for years. Tip number eight: Letterbox your spot.

Hopefully, the rest of the things I'll mention will be common sense. Using an analog tape that has 600 plays will be full of dropouts. The viewer doesn't know what that is, but they know it looks ugly. After the spot is edited, look at it. Most dropouts don't develop until after the tape has been played too many times. If you don't notice it at the beginning, it won't show up later. Tip number nine: Dropouts are evil (this isn't a problem with digital).

Give your screaming narrator the day off. Ask the client if Aunt Sophie can sit this spot out. If the client really prefers their aunt over one of your narrators, you're stuck. Doing a voice-over is a special gift. Some can read really well but have a poor voice. Some have fabulous pipes and can't read. Test your possibles for both. Train them if promising, but you know what they should sound like. Tip number 10: If the voice over sounds bad, the viewer will notice too.

I've discussed some of the basics in making a spot look "not local." You know what national spots look like. See what they do differently and try it yourself. Don't copy; try your version of it. Let me know if it works out. Tip number 11: Read all the tips and try them. You might amaze your toughest critic.

All my commercials have a disclaimer that states, "the following commercial is a national spot; it was just produced locally." It works some of the time.

How to Make a Commercial with No Camera and No Budget

A lot of times we're given budgets that are small enough to only be seen with a magnifying glass, but we still have to come up with creative ideas and a spot that will make the world a better place. I had three such clients that had grand ideas but very little money to spend.

Holiday Spot

The first spot was a 10-second Columbus Day craft show announcement. Tons of artists from all over the planet were going to converge on this four-acre park over the Columbus Day weekend. As you probably know, you can't do or say a lot in 10 seconds, but once again this client wanted the world. In 10 seconds, he wanted to tell about all the juried craft people that would be attending, where it was, when it was, how much it cost, what number annual event it was, what color his wife's socks were, and some pretty memorable voice-over narration.

Because this client only had 10 seconds to deal with, this was going to be a jam-packed spot. It had to be full of character generated (CG) text to explain the times, dates, locations, and so on, and it also had to be interesting. This is the point that he kept on making as he poked my chest with his index finger.

When the time came to talk about money, he seemed to have very little to say, and even less money. In fact, he had less than $100. Of course, this was just the money he had to spend for production costs; he would probably end up paying 50 times that amount in air time costs.

To make the client happy, I came up with an inexpensive (I didn't say cheap) approach that hopefully could make everyone happy. I found a picture of the *Santa Maria* (it's one of Columbus' ships, not the southern California town) in an elementary school history book. As visions of sharing a cell with, that's right, Snake, danced in my head, I contacted the publishing company and asked them if I could use the drawing in a commercial. I had to use the drawing instead of a photo because the photographer was probably sick that day (back in 1492). When they found out my budget, they said I should have saved the long-distance phone call. I had permission; rather rudely put, but at least I had it. In Figure 3-1, Columbus never looked so good.

Our graphic artist friend, Steven Brehm, electronically cut out around the drawing expertly and supered it over zero black (super black). He then

Figure 3-1
Columbus' ship

took a cloud background from one of his photo clip files and created another image. On his system he was able to accomplish these two feats in a matter of minutes. If I had to create the images myself, it would have taken a better part of a week to get the machine turned on.

In the *nonlinear editing* (NLE) system, I created a 10-second clip of the clouds. I found another clip of a deep, blue ocean and layered that on video channel 3. By softening the edge where the clouds met the sea, I had my background.

Now that my background had been created, I just had to make the ship move. With the image of the *Santa Maria* still in the editor, I simply created a motion path to have the ship move from right to left on the surface of the water and get smaller as it sails off into the horizon. This movement took quite a lot of practice so it would look convincing, instead of just cheap. The CG text was added and totally filled the screen.

It still was a 10-second commercial, but the client was happy (that's the reason we were put on this planet) and the spot had an interesting, storybook feel.

Web Site

The next project wasn't quite as easy in design as it was to create. This client created web sites but didn't actually want to show any of his web sites in the commercial. Because he was starting out, he also didn't have any money for production. Yours truly, Captain Cheap, was called into service. I came up with the concept of trying to leave as much to the viewer's imagination as possible. Since I couldn't show any web sites, I had to do the next best thing: have the viewer imagine one by using the simple concept of audio. Audio would have to carry this commercial, not the visuals. Basically, this was going to be a radio spot on television. In fact, after we had completed the spot, the client turned around and used the exact same spot on the radio. Twice the bang for his buck, which was almost half his budget. In Figure 3-2, this cheap spot could be read and heard.

The stereo spot started out with office sounds over a black screen. A woman then complained to her manager that she wasn't able to get the work done because she didn't have access to the Internet. A narrator then came on and explained how easy it was now to have Internet access.

All the visuals were going to be CGs. These CGs would tell what the client could do for the viewer while his URL logo was displayed in the lower right.

Now in this type of spot, it's important that the actors/voice-over people sound convincing like they are actually having a conversation instead of trying to sell you a used Bronco. The visuals were just there to give you the

Figure 3-2
Web site spot

pertinent information, and they also mimicked what the narrator was saying. So the inexpensive way to make this spot effective was to create a radio spot for television. I'm not saying that radio spots are cheap or ineffective, but they usually don't work on TV without visuals.

My First Time in Jewelry

This last example I'll be talking about is the really, really, inexpensive spot. You really don't even need a camera for this type of advertising.

The client was having a jewelry sale and didn't want to spend a lot of money. He quickly produced a color catalog that had every piece of jewelry in it with its correct sale price. It might have been more cost effective if he just mailed the catalog to everyone, but he wanted it on TV. Floating images and great prices make this jewelry spot shine, as shown in Figure 3-3.

We came up with a spot called "Quotes." Six images were selected from the catalog, scanned into the computer, and saved in Adobe Photoshop. The pictures weren't enhanced in any way, shape, or form. While still in Photoshop, text was typed that said things like "unique styling," "affordable prices," "large inventory." I think you get the idea. These images were assembled onto an NLE and with groovy music; we had a spot.

Each of the concepts I mentioned were among the dozens of spots that most local TV stations must produce for clients with minimal budgets. The photos used to illustrate this piece are actual still images from the completed video. None of these commercials are really time consuming or really that creative for that matter. You just have to use the resources available to

Figure 3-3
Jewelry spot

you at the time. At no time was a camera used for these spots. Maybe we should sell our expensive camera and just do inexpensive spots from now on. No, I get more attention with a camera attached to my eye.

Review: Five Things You Can Do with No Money

Here are some alternatives for when you have no budget to speak of:

1. Look through old books or clip art files to find the images that will entice as well as suit the bill. For very little money, you can purchase stock photo art on CD and have images without taking the time to shoot them (just stay away from your vacation photos at Dollyworld).

2. Don't be afraid to layer your images. Try to use four or more photos to make up one shot if a lone image doesn't seem to work. It may take a little longer to render, but you now have a combined image no one else does. You always wanted people to say your work was deep, and if you have eight images together, you can now say, "it's deep."

3. If visuals can't or won't tell the story, try using sound images. Even on television, the sound can carry the spot as well as the visuals. You should show something on the screen, but let the sound do most of the work. If shooting a commercial in a snowstorm during a whiteout, your sound will have to work twice as hard.

4. There's a very effective *public service announcement* (PSA) for all the heroes (police, firefighters, and so on) who risk their lives daily for our safety. The spot has *no* audio, and the graphics tell the viewer to have a moment of silence for these brave people. Here the visuals carry the story and the silence makes this PSA stand out. If you want your message to stand out, let the audio or video alone carry the spot without the other. You will be breaking tradition, but sometimes that's just what the doctor ordered (mine's usually chasing his nurse).

5. Use movement to catch the viewer's eye. A camera is seldom needed if you want smooth, flowing images (with no budget). Have the images (art or text) slowly move across the screen. Normally keep the movement left to right unless you want to jar the viewer. Anything that flows against the normal grain will attract attention; just make sure you want that attention. If you're married, just say no.

I've Got the Chroma Key Blues

One of the exciting things about chroma key is that you can create something that really isn't there. You don't have to travel to the Swiss Alps to have a great shot of snow-capped mountains. It isn't as if you're lying or doing something phony; it's more like you're enhancing the shot, and that's a good thing.

As you probably know, a chroma key background enables you to super just about any image on its surface. It's used every day in weather reports, music videos, and commercials. Although many different types of chroma keys exist, such as chroma key blue, chroma key green, and, the most effective, Ultimate, it's now as easy as putting the background on one layer and the image on another.

The concept of lighting all of them is basically the same. The example I'll be talking about is chroma key using blue screen. But as I said earlier, you can apply the same lighting techniques with any form of chroma key. The stark contrast of chroma key blue can be seen in Figure 3-4.

Our goal on this particular shoot was to have a two-person, over-the-shoulder interview using a blue screen behind the talent (a boring commercial, but that's what the client wanted). This is the same type of setup that *20/20* and *Dateline* use on their interviews (only without the blue

Figure 3-4
Chroma key
background on a set

screen). Since Barbara Walters wouldn't allow us into her house for the shoot, we had to rely on chroma key.

On this particular blue screen, various background images would be substituted over the blue image. This concept helped Superman fly, allowed Dad to shrink his kids, and kept Godzilla in the dark.

I have been told the reason this color blue was chosen was because it didn't appear that frequently in nature (have you ever seen a chroma key blue lion?). I guess they really meant that most styles of clothing didn't use that shade or hue of blue. It's a very vibrant color. Our talent chose to wear a light blue shirt for this shoot, but luckily we had no trouble with the background images appearing in his shirt.

I had always heard horror stories at the TV station where the "new guy" would be ready to do his first weather report on the air and had a low pressure over Kansas appearing in his tie. Or the poor unfortunate woman whose contact lenses happened to be the exact same shade as the blue screen. Her eyes really told the whole story. But for the most part, this particular shade of blue is unique and its chroma content is quite saturated. I believe that this is the reason that our talent's blue shirt wasn't a problem. It was far too light.

To create our set for the interview shoot, we stretched a sheet of muslin tightly across a 2 by 4 wooden frame. The frame's size was 8 feet by 14 feet but was assembled using 2 by 4 lumber. Once the muslin had been tightly stretched and fastened, it was painted with chroma key blue paint.

The talent's chair was eight feet in front of our blue screen. A small wooden box (our coffee table) was to the immediate right of the chair. If we had covered the box in blue, the background image would have appeared in that space. We didn't want the interview to look as if it was happening in a void. By framing the talent in a medium shot, no one could tell that we were using a blue screen at all.

This brings up an interesting point; do you really want the world to know you're using a blue screen or would you like it to remain a secret? The telltale blue fringing around the talent's hair and clothing is no longer a problem with the sophisticated units on the market. We've come a long way since Tiny Tim (the Dickens character, not the late singer).

George Winchell, our lighting guru, created the lighting pattern for this set. The main thing to remember about chroma key is to light the blue background as evenly as possible. Any shadows, light spots, or wrinkles might not key as well in editing, and that would definitely give our secret away. Because we were shooting this interview and adding the backgrounds later in editing, we did several tests with a light meter to make sure the background was evenly lit.

Using two lights, one on the left and the other on the right, we had an even F4 on the background. Our 2K open-faced lights had to be diffused with tough spun to soften the blow on the blue screen. These newly created softlights' only purpose was to illuminate the background. The talent needed lights of their own. Placed at a 45-degree angle to the background, each light was 6 feet high and pointed at the blue screen. By moving each light carefully and playing with the barn doors, we had an evenly lit sea of blue.

Our softlights came into play for the talent. Each one of these 2Ks was used as a key light for the talent, one for the interviewer and the other for the interviewee. Once diffused with tough frost, the talent had an even, soft highlight on their faces. In fact, some of the interviewer's key light acted as a fill light for the interviewee and vice versa. The softlights were placed high enough so the shadows they cast would fall on the floor rather than the blue background. Remember, in love, war, and chroma key, everything must be even.

By using a net in front of the softlight, the talent's face would appear less washed out and have more detail. You're free to light and model your talent any way you like using conventional lighting; the only trick about chroma key is to keep the background free from shadows. This shoot would have been lit like any other interview setup; we just had to make considerations for our chroma key.

In post, our blue background was removed and our nifty mountains were added. Of course, some people could have traveled to the Alps, but we like to live on the edge (of the mountain) whenever possible (and we are talking no budget here).

Review: Five Things to Remember When Trying to be Blue

The main tips to remember when using a chroma key background are as follows:

1. Even lighting is the key. Any shadows will not disappear as well as the lit color. Flat, soft lighting will produce less shadows than an open-faced light or Fresnel. If you don't have access to a softlight, try bouncing or diffusing your source. Remember, you can get rid of shadows by adding more light; you just add another shadow.

2. If you don't have access to blue, find another solid color. Blue is nifty because most clothing doesn't match that particular hue. NLE systems will enable you to "key out" any solid color. Just making sure your talent isn't wearing that same one (your turn, insert joke here).

3. Everything you want to disappear or key out must be the same color. Don't mix blue and green as your chroma key colors. Besides looking strange on the set, your computer system won't know what to key out. Try chroma key paint or fabric as your mobile backdrop. Remember to take it with you at Thanksgiving; Uncle Rodney always looked better in the mountains.

4. For those of you on a low budget and want the look of a virtual set, shoot the keyable scene with a cyc. A cyc is basically a room with all the visible surfaces the same color with the walls curving to meet the floor. This can be built with plywood and paint. The computer will then make the solid color disappear and your actor is in a virtual set that cost virtually nothing.

5. Use keying sparingly. Most lower-budget systems leave a slightly perceptible telltale jaggy line around the edges of a keyed object. If you can't afford to shoot in the background, the subject can't fit, or radiation poses a danger, use chroma key and you'll never be blue.

Don't Call Us, We'll Call You

Sometimes clients really want to take their advertising to the edge. What I'm getting at is, if their concept really sticks in the viewer's mind and they don't offend too many people, they have a winner. If a commercial isn't memorable and people forget it moments after they view it, it was unsuccessful.

A corporate marketing guru of a local mini-market, one of 60 throughout the state, wanted to create a campaign that would tell the viewers that their old stodgy image was changing and that they were on the cutting edge. They wanted to look ahead; the old was left behind. In addition to this new style of commercial, they still intended to highlight three different products that would be on sale that month. Their old spots involved a long shot of the store's exterior, three products (that were on sale) on a table, prices in the lower right-hand corner, and voice-over narration, and with the spot closing on another exterior shot. Somewhat effective, but everyone else does it the same way because it's inexpensive.

It was decided that the best approach for this campaign would be to create a donut commercial. The beginning and ending of the spot would remain standard and the middle (the part with the products) would change every month. This was a simple concept that had been done a million times before, but our version was going to be different.

The marketing guy had approved the storyboard and concepts and wanted to get under way as soon as possible. I guess that's why he was taking his blood pressure every five minutes. Not to stereotype, but he was the typical new head of marketing. The old head of marketing was 105 and had to be replaced with new blood. This guy couldn't have been out of college more than 12 minutes. His first day in office he wanted to change everything that had ever been done before. It would be his new way or no way.

Our concept was simple. We decided to create a fake talent search for the next spokesperson for the AM/FM mini-mart chain. Everybody in the world goes to a mini-mart to get milk, cigarettes, candy, gas, or something. What better way to tell the world about your new image than to do it on TV with an on-camera spokesperson? Of course, this wouldn't be some attractive, likable male or female. It had to be someone with a "bit" of special talent. Mr. Ed (our favorite talking horse) would have worked nicely.

To take it one step further, having a nationwide talent search would peak the interest of the viewer at home. Even though this was only a statewide chain, if we said we were looking all over the country for the right person, we would spark more interest. These potential new spokespeople might even wander into one of the local mini-marts and ask who to see to audition for the screen test. This is advertising at its most effective and was all part of the game plan. In fact, this even got our marketing guy excited (it was either our great idea or the fact that I kept dropping money out of my pocket every time I sat down).

Even though we were shooting in video, we wanted to give the impression that we were shooting a screen test on film as part of our talent search. The open graphic in our donut spot was to have the words "the search is on" inside a marquee-style billboard, much like something you would see at a movie theater. With the words in the center, the marquee lights would chase around the perimeter. This title really was up in lights. At the bottom of the frame, two search lights were criss-crossing. We wanted this to look like a gala world premiere event. The marquee and search lights were supered over the door of our imaginary studio. Actually, it was the door to a broom closet, but the narrow little door added to the humor of the spot. When people would walk in to audition, they would open this tiny broom closet door and audition in a 4-by-4-foot space. We wanted it to appear glamorous with the marquee and search lights, but also wanted it to look tacky with a minuscule auditioning room.

For the three and a half seconds that this image filled the screen, a booming baritone voice (à la Ed McMahan in the old *Star Search* series) would read the words the marquee highlighted. This image then immediately cut to a close-up of the clapboard. The slate had the name of the mini-mart, the talent search title, the date, and Take #346. We wanted the viewer to believe that we had been auditioning for months just to find the right person. The top of the slate was open and immediately closed with a "clap." Although you never need this synching device in video, we just wanted it for the effect. After this one-second slate shot, three frames of white were inserted with a one-kilohertz tone. This three-frame tone becomes a "beep" in the finished spot. Once again, this fogging white and beep imitated the frame fogger and tone that a sync film camera creates. Now at least we had something to sync the picture and the sound to. The overall effect was to look as if we pieced together all these bits of talent searchees, opening slates and all. When some new star becomes famous, they always just happen to find an old screen test of them somewhere, complete with the opening slate.

Although we didn't have time in 30 seconds to have the door of the broom closet open to reveal our auditioning space, we cut from the closed slate to our first screen test individual. In order to make our broom closet look like a real TV studio, we actually rented space in a TV studio. The backdrop was a white curtain that acted as a cyc that surrounded the perimeter of the studio. One light was gelled red and extended to its maximum height. The barn doors were closed so that a streak of red light would fall across the white curtain. Another light was set up on the left side, this one with a blue gel. With our red- and blue-streaked curtain illuminated, we now had our backdrop. In order to more fully dress our faux set, a plastic plant, a wooden stepladder, and a video monitor were added with the camera's frame. When the talent auditioned, we would actually see what the camera saw in the onscreen monitor. How's that for immediate results? In Figure 3-5, this search candidate has money to burn.

Using a five to one key-to-fill ratio, I wanted the talent to be lit extremely well on the left and just a little fill on the right. While keeping the studio effect, I still wanted it to look like this was something that was done in a hurry. It had to look cheesy even though every light and shadow was exactly where we wanted it.

To add still further to this screen test set, two Fresnels were lowered so that they would appear in the frame. Although they weren't turned on, the look of two large, gray, barn-door lighting instruments added to our set. The world had to believe we were doing this in a real studio (or broom closet). A director, complete with director's chair, baton, and beret was seated off to the left. His silhouette would be seen in the frame. Because the talent's

Figure 3-5
Search candidate

kicker light was backlighting him, the director's profile made him look like Orson Wells overseeing an important scene. Even though you'd never see his face, his imposing form and wild gestures would be enough to get his thoughts across to the viewer.

Each screen testee would be shot in a medium close-up to a long shot, depending on what they did. The first victim was someone who said, "If you have money to burn, go someplace else." On the magic word "burn," he opened his wallet and his money burst into flames. In order to make this as effective as possible, he was framed in a medium close-up, and the sound of the flames engulfing his wallet was enhanced in postproduction (crinkling cellophane). The flames shot a full 10 inches from the wallet and looked like one of the trick cigarette lighter flames that Jerry Lewis always used, setting his nose hair on fire. When the wallet started to burn, the director jumped back in his chair and shielded his face from the flames. Obviously, we staged everything that happened, but we wanted it to look like this director had no idea who was coming in to audition next or what they were going to do. His off-camera voice would yell "cut" after each person completed their act. Each person's act would get stranger and stranger as the auditions wore on. The second interviewee in Figure 3-6 was a monkey's uncle.

After the wallet burned for a few seconds, I cut to the snapping slate again followed by our white flash and beep. Our second interviewee is difficult to describe on paper; he worked much more effectively visually. A black

Figure 3-6
Second interviewee

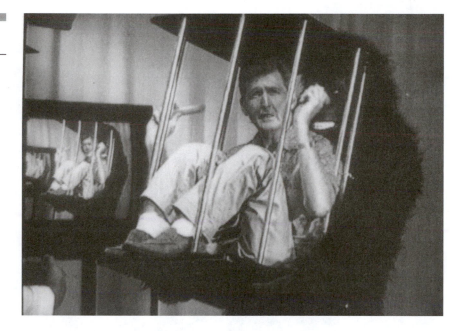

gorilla walked onto the set in a long shot, carrying a man in a cage. Without giving away any secrets, the man's legs and torso would be inside the gorilla suit and his human head exposed in the cage. The rest of the human figure in the cage would be a dummy. The man's hands would also be free and visible and allow him to grab onto the bars of the cage or to throw bananas. When the gorilla and caged man walked onto the set, the director jumped up onto his chair thinking this was a wild animal. This had to be done correctly or it would really look hokey. It was necessary to shoot this in a wide shot so the viewer could see the gorilla walking in. Although the gorilla looked like a reject from an old Three Stooges movie, it added to the humor. As the man in the cage began to do his pitch, the director was shaking his head, covering his eyes, and yelling cut; vaudeville had returned. Every reject from a circus would be auditioning for the mini-mart spokesperson.

At this point in the spot, after the two auditionees, we dissolved into the first of three products. Each product was placed against a neutral background (with soft, diffused, colored lights) and appeared on the screen for five seconds. These products would be changing every month. A voice-over narrator would then talk about the products and their prices.

After the third product, the slate would appear for the last time. On each month's commercial, the last talent searchee would be the most outlandish, or the stinger. Our third person was an Elvis impersonator, as shown in

Figure 3-7
Elvis impersonator

Figure 3-7. He was clearly a hunka, hunka burning talent. Complete with white-sequined jumpsuit, black pompadour hair, sideburns, and guitar, this thinner version of Elvis gyrated in front of the camera. As he sang "you ain't nuthin' but a hot dog, mustard on the side" (to the tune of "Hound Dog"), his hips swiveled. He grabbed a dummy mike stand and dipped it as he would a dancer. As he was singing, the director covered his face yelling, "CUT, CUT!" If only the director had listened to his mother and gone into a respectable position, like bookmaking. If only there weren't so many talented people who had to audition for this role. If only Elvis had left the building.

The last image on the screen was the marquee again with "The Search Continues . . ." keyed over the broom closet door. Elvis's voice is heard over this last shot saying, "thank you, thank you very much" in his own impeccable style.

Throughout this 30-second spot, "loony tunes" music is heard. This comedy track pokes even more fun into this far-out idea. Although very tightly edited, this spot showcases three hopefuls trying to land the role of spokesperson.

Each month's spot will feature three new search hopefuls such as the three-handed man, the Swami, Guido the Human Turnip, and many more. Ideally, the very last commercial at the end of the run will have a winner to our star search competition. Much like *The Fugitive* TV series, we will shoot an ending commercial with a definite winner before we complete the run. If this spot is already in the can, we can use it at any time. If the spots run 12 months, we'll shoot a winner early on and use that when the client tires of us. If it runs 10 years, we'll still have an ending spot to use at that time.

All in all, this type of concept is an incredible leap of faith for any client. This is not the stuff people usually see on television. The viewing public has gotten more refined and discerning in their tastes. It's difficult to try something new (not what the viewer expects) and have it be accepted. It must be planned out, worked, and reworked. Our concept was strange, unusual, out of the ordinary, and downright silly. It will get people's attention. If it gets enough attention to get people to come into the store and try to sign up, talk about the spots in their sleep, develop fan clubs and cults, or get into advertising, then we were successful. If none of this happens, then the director and I will audition for the spokesperson's role. I'm pretty good with a screw gun . . .

The budget for this fiasco was a measly $500 per spot. The "talent" worked free for the exposure and our only cost was the rental of space and equipment.

Review: Ten Reasons to Prove You're in Charge

The keys to remember here are as follows:

1. When creating a donut (get up early to make them), make sure you know how you're going to fill the hole each week. The interior must be the same time each week. If possible, leave yourself a little leeway with the open and close, allowing more time for the hole, which is really the "meat" or most important part of the spot. Without this hole you would have nothing to sell (that's why they sell donut holes).

2. Come up with an interesting concept because the viewer may be subjected to this week after week. If the opening and close hold interest, then the inside or "hole" will be worth watching. A "continuing story," "where can I find . . . ," or "this week

(continued)

(continued)

at . . . " will pull the audience into the black hole of your donut (a chocolate donut, what did you think I meant?)

3. Make the audience think you have a bigger budget than you actually do. Most sets on TV are cardboard when seen in real life, but the viewer will never know that. If doing a store spot, use graphics or cardboard to enhance what's really there. You aren't really lying, just stretching the truth a whole lot.

4. What the viewer can't see, they have no clue exists. Only show what needs to be shown; use your imagination. By keeping things out of frame, enhancing sound effects, and being slightly unusual, you should have the viewer where you want them because the client is after their money, and your job is to brainwash them into letting you have their money.

5. If using characters, make them likeable, believable, and outlandish. You have to cut through all the junk on TV and make your spot stand out. The only way to effectively do that is to be unique. Don't insult, talk down to, or belittle the viewer. Entertain them in a unique way; then ask for their money.

6. Be consistent week after week with your donut and people will want to see what happens next. If they like your characters, they will tell their friends about "that new commercial." If something changes one week in your donut that shouldn't, it ceases to become edible and is now stale.

7. Be consistent, but also change only slightly. With a donut, some things remain the same (like the song by Led Zeppelin), but slight variation is good. The sale products will obviously change, but keep the rest of the spot fresh with your approach.

8. Because you may be doing this spot week after week, make the consistent things easy to shoot. Keep all the products in the same area with the same backdrop so changes will be minimal. If you know what the future weeks' sale items are, shoot them at the same time as the previous weeks. That's why everybody likes donuts; just wash your hands after eating a glazed one.

9. Leave the viewer wanting to watch the next installment. If the spot is dull and lifeless, no one is going to care what happens next week. Build upon last week and make next week's even better. The client doesn't really care about this; it's not their job. But it's your job to make the spot interesting. Having donuts on the set makes shooting a donut more fun.

10. You may not often get a chance to be creative in a donut. But by doing so, you can make the boring exciting and ask the client if it would be something they would want to watch week after week. Whatever they say, have fun with it anyway.

How to Make a 35mm Commercial on a Video Budget

Most of the commercials you produce will be on video, but once in a while along comes a client with a real budget who wants to shoot in film.

Clients have very little money to spend on the actual production of their commercials. Most of their money seems to go for the air time. In a sense I can see why. Someone could make the best commercial possible, but if it gets no air play, what good is it? It really doesn't cost that much more to shoot a spot in film than it does in video if it's done correctly. Now before everyone starts yelling at me and telling me that I don't know what I'm talking about, let me explain. I'm talking about a highly produced, scripted, shot, and edited spot. I've had more than my share of $50 spots. Unfortunately, most of them ended up looking like they cost that much. But if a little more money *is* there and you have time to plan ahead in scripting and shooting, you will have a spot that looks like a million, but only costs thousands.

Each week at my TV station (I don't own it; I just worked there), at least five commercials had to be produced and completed. Some days we had three to complete in one day; other times we had a week just to do one. One Monday afternoon a client came in and wanted to do a food spot to entice viewers into his local establishment. Before we even talked money (they always seem to bring that up first), he told me this spot had to look good. His restaurant was known for the presentation of its food, which looked good (it might have tasted like cardboard, but that wasn't the issue). Each week the local coupon-clipper magazine would feature color ads of these amazing looking steaks, lobsters, crab meat, and pasta. The client said he wanted his restaurant's food to look as good on TV as it did in the coupon ads. He also said that he didn't want his food to look like the yellow or green Chinese restaurant food that he had always seen on television. His food had to look real, not pasty.

I explained to him that his coupon ads were shot on medium- or large-format film, and that video couldn't come near to that resolution. I told him that we would make his food look extremely eatable with great lighting, gels, and broadcast-quality equipment, but in all honesty it wouldn't look as great as it did in the ads.

As he began heading out the door, I told him the only way it would look the way he wanted was to shoot the spot in film, not video. As he closed the door, he said, "Fine, shoot it in film." I guess that was that. Of course, it's impossible to shoot a spot in film and make it cost the same as video, unless your last name is Eastman Kodak or Fuji. There are just more expenses involved.

So now I had the client's permission to shoot in film and I didn't even get a chance to tell him how much it was going to cost him. I had to tell him eventually, and after he sold his yacht to pay for it, he was happy. Hopefully, you can learn from my mistakes and save even more money.

Since I didn't have a film camera lying around at the station, I called rental houses and checked prices. If I could shoot the spot on a Saturday, I could add more time to the shoot without paying extra for the camera. The camera would come in on a Friday; I'd shoot on Saturday and Sunday (that makes your crew very happy) and send the equipment back on Monday. I really had the camera for three days and paid for one. Thirty-five millimeter looks better than 16mm because the negative is twice the size, but camera rental costs are very close. Sixteen millimeter is cheaper because the film stock costs less. Figure 3-8 showcases a rented 35mm camera setup.

Also, don't be afraid to tell the rental house that you work for a TV station (don't tell them you do if you don't; that would be lying). Tell them what your budget really is. They may even take pity on you and give you a price break. At least they did in my case. The station wasn't really making that

Figure 3-8
Film equipment
on a shoot

much on the production costs, and it was a great chance for the crew to work on a film shoot. I had experience in film production, and the rental people were very nice people. They'd like to have the business, they want to make you happy, and it doesn't hurt to ask. Just make sure you have a grand total for all equipment, shipping costs, and insurance. Your client doesn't like hidden fees or charges any more than you do. I got all these costs down to $400 (not bad for a $60,000 camera).

Don't try to use video in your film camera. You can buy brand-new film in large lots and save some money. Or you can purchase what are called *short ends*. These are odd lengths of film that are left over from larger shoots. It's still unused film, but whoever purchased it only needed part of it and returned the rest. You can really save a bundle with short ends. We were able to get 390-foot loads instead of sealed 400-foot loads and paid half the price.

One core of film costs about $250. Our shoot only called for two cores (that's roughly eight minutes of film). In planning the amount of film you need, it's best to shoot as tightly as possible. This is where a shooting script and storyboard really come in handy. Rehearse everything before you shoot, and practice your camera moves before you roll film. Film stock is *expensive*. Buy only what you think you'll need. I know that's tough, but that's what they make dart boards for.

The last, most expense part of the film experience is getting the film processed. It's very inexpensive to develop the negative, usually only about $.15 a foot. Now you have two options, one's expensive and one's very expensive. Once you have the negative developed, you must either print it on positive film or transfer it to video. Unless you have access to a flatbed editor or film-cutting equipment, it's far easier to transfer the images directly from the negative onto videotape. Most film processors offer this service. Call ahead and find out their best rates for video transfer. Night rates are less than day rates, unsupervised are less than supervised, and you can save some money if you supply your own tape stock. I won't go into the many beauties of color correction (that would take weeks), but once you get your tape back, you begin the editing process and your costs remain the same. The only difference is you now have a film master on video.

As you can see, besides the cost of rental equipment, film stock, processing, and transfer, film shoots are not that much different than video shoots. If you plan ahead, script well, call around, and get the best prices, you will have a finished product that will knock the socks off the competition, assuming they wear socks.

Review: Five Reasons Why Film Is Worth the Expense

In case you're not already convinced, here are the five reasons:

1. Using a film camera rather than video will increase the expense but make the final product look better. You must justify this cost to the client. Have them look at national commercials with the same type of product and point out their film origination. If you can't sell them on the look, they won't pay the extra money for the camera. It is cheaper to use older equipment, but your job will be slightly harder. In order to save money, I borrowed Charlie Chaplin's hand-cranked camera (he wasn't using it).

2. Try to get a deal through a rental house. If you have no budget, tell them and they may help you out. When you walk in, have your pockets turned inside out.

3. Shoot over a weekend if possible to get an extra day of camera usage. Most rental houses charge four days for seven days of use. This sounds like bad math, but it may work to your advantage. The same applies to holidays, but you may have to pay your crew extra.

4. Use re-cans or short ends to save money on film stock. If you want to break the shrink wrap on a new core, you will definitely pay for it. Each second the camera is running you will hear the money disappearing. Rehearse and practice before the camera rolls.

5. The cameras look bigger and make you feel more important. You can actually tell people you are "filming a commercial" because you are using film, and your captured image will be sharper than video. Just remember to use this for good rather than evil.

Making a Car Commercial with Someone Else's Footage

It's not stealing. In fact, that's what they want you to do and that's why they shot the footage in the first place. If you work at a television station and you make car commercials, you know what I'm talking about.

I'm one of those old-fashioned people who look forward to the new model year in automobiles (the fins of the '50s are coming back). So obviously I'm very excited when I get that new tape with the latest model cars.

I've made numerous local car commercials over the years. For many of these, I've traveled down to the dealership with my crew and we'd shoot something there in the showroom. This works well with used cars and specific incentives that a particular dealership might have. But there's something about the footage that's used in the tapes you get from General Motors, Toyota, Ford, and so on. Most of this footage is shot on 35mm film with stabilizer mounts, helicopter shots, Steadicams, and exotic locals.

First of all, how do you get this footage if your local car dealer requests it? Somehow they always seem to know that the footage exists somewhere, but they never tell you how to get it. While working at an NBC affiliate, the car people who made these classic tapes found me. Because our station aired national, big-budget car commercials, they knew we had to have local dealers chomping at the bit. Usually around July they'd send a letter to the Commercial Department. If you got a letter like this at home, it would have Dick Clark's picture on it and say that you have already won lots of money. But if you get it on the job, it's a different story. In this packet would be a postcard that asked if you wanted Betacam SP, and now DVD. Some of these tapes were free (Chevrolet and Cadillac) and others wanted a nominal fee (Lincoln Mercury—$100). These fees basically just paid for the tape footage and shipping, and the tapes were worth every penny for the footage you would get. In all cases, you were allowed to use as much of the footage as you wanted and could air it as often as you liked.

If you don't happen to work at a TV station and still make car commercials, I suggest that you contact the local stations in your area and ask if you could use their footage. Before I worked at my station, they had miles (pardon the pun) of this footage that they got free every year and never used. It doesn't hurt to ask and all they can do is shoot you. Besides, the footage is usually worthless when it's a year old (unless it's BMW).

These tapes are sort of like a PSA for the dealers. The footage always looks phenomenal, every model and color is represented, and by using this stuff you'd be making them look good. They shot all this footage for a reason: to be used (that wasn't meant to be a used car joke).

To give you some specific examples, our local BMW dealer wanted a local spot touting a great sale they were having on their 330i series convertibles. I asked when would be a good time to come down and get some footage of these cars in their natural environment (in my driveway). They said that they didn't actually have any of these vehicles in stock. So here was this dealership wanting me to shoot a commercial for cars that would be on sale that they didn't have! They also said that it wouldn't be a problem not having any cars; since I was in video, I would have that footage on hand. My wife always lets me collect video footage on every new car marketed. (Right!)

I went home and found out I was fresh out of 2001 BMW footage. I told them I did have some old film footage of the 1966 Chrysler Imperial that I had recently come across, but they found no humor in my statement. I also told them that I didn't know where I could get that type of footage within a week, but since they were a dealership, they may have more clout with the Bavarian Motor Works.

Three days later they called me and had a Betacam tape with the 2000 models. I said that I thought they were selling 2001 models. They said they were, but the cars hadn't changed. At least the 330i ragtops had not.

To sum it all up, I now realize that this high-end dealership never would have been happy with the footage I would have provided on their limited budget and time frame. These 330s on the tape I received were driving through the Swiss Alps with perfect people behind the wheel. If they had wanted the 330 behind an Amish buggy, I was their man. In Figure 3-9, beautiful cars still exist.

This pristine footage featured moving camera and locations that you only dream about, without a speck of dust or dirt anywhere on the vehicle, and had four different colors to choose from (cars, that is).

Since I now had this great 35mm footage, all I had to do was edit it together. I came up with something creative, cut it to the beat of the music, had an upbeat voice-over recorded, and inserted their CGs detailing price, leasing information, and locations. All in all, it was an easy spot to produce, it looked like it cost big bucks, and the client was happy. Remember, we're

Figure 3-9
Car advertisement

talking about an expensive automobile. In lieu of payment, ask them if you could have one of the cars for a . . . day?

Apple Pie and Chevrolet

Flash forward to a Chevrolet dealer. I cold-called them and told them I had just received 60 minutes of the newest model year cars and asked if they wanted a commercial made. When they found out that I didn't have to shoot any footage (thus lowering the cost), I could update it throughout the year with new sale or lease prices; they jumped at the opportunity.

In fact, if I was really devious I could sell the exact same commercial to all four dealerships in the area and just change the CGs and logos. But since I had scruples (not the board game) I decided against it.

Even though the vehicles in the Chevy tape were less expensive than the BMWs, the footage looked just as good and offered as many variations. There must be a place somewhere out West where beautiful people, flawless cars, and stupendous locations all exist. No matter how you edit these spots, they all seem to look great because you have perfectly exposed footage.

Our local dealer seemed to like sound effects. During editing, the images would bounce onto the screen (via sound effects) and slide on and off. I could use almost every NLE effect I could find because that seems to be the way car spots are edited nowadays.

Unless you edit everything upside down, the spots will get the viewer's eye (upside down will also get the viewer's eye). That's precisely why these tapes were produced: to be used in a manner the dealership deems fit (no royalties), to save all that wear and tear on your body having to travel to far off places (with strange sounding names), and to make you look good in the local dealer's eyes. I think making money enters into the equation someplace, but I'm not so sure.

▮ Review: Five Reasons Why Someone Else's Footage Works for You

1. Ask local dealerships or TV stations for car footage. Most times it's sitting on their shelves unused. Just make sure it's current (look at the year; if it's before World War II, it may not apply). This footage does have an expiration date.

(continued)

(continued)

2. Use your editing prowess to make the borrowed footage sing. The originals were high-budget productions and your editing will have the look of a national spot.

3. You now have access to exotic locations, beautiful models, and sometimes rare cars. These are great spots to put on your demo reel. Don't say you shot the footage if you didn't, but great editing can make good footage look even better.

4. Using preexisting footage keeps your budget down. Minus a small acquisition fee for finding the material, your only costs are in putting the spot together. Sticky glue doesn't work here.

5. Contact every car dealership in the area once you've obtained footage to present your pitch. Who could possibly turn down an offer like that? Most will, but it was fun asking.

How to Write a Car Commercial

The biggest drawback about a 30-second commercial is that it only lasts half a minute and that's not really a lot of time to tell much of a story. When the public views the completed spot on the air, they never realize that you told them the when, where, why, and how much in less than half a minute.

Several things go through my mind when I'm trying to write an idea for a commercial. I've met with the client, I found out all I could about the product, and now I have to dive in head first.

The particular case I'll be writing about was a spot for a local Ford dealership. This dealership was known for their screaming announcer demanding that if you want the best deal on a new or used car, you'd have to visit their dealership. I had heard that the public tolerated the obnoxious ads but weren't coming into the dealership in droves because they were offended by the hard-sell tactics of this "screamer."

Upon meeting with the client, he agreed that we should change this approach slightly and still do something unusual but not offend the public by screaming at them. The name of the dealership was very recognizable in the area and whenever I mentioned their name, people would always say, "Oh, that's the place with that screaming guy in the plaid sports jacket." I really had my work cut out for me.

In my mind, humor usually sells best. People will remember a commercial a lot longer if there's something funny about it. Along the same lines, people will also remember a spot if something out of the ordinary happens.

How many car commercials do you remember when you just see various new models driving by? If one of those same cars morphs into something else, climbs up the vertical side of a building, or becomes dust and vanishes, that sticks in your mind.

In devising a new game plan, I decided to use humor rather than hard-sell tactics to change the public's opinion of this place. The new spot I had written centered on the Ford name. The spot would open with a man dressed in period clothing tugging on the reins of a horse (no, it wasn't a Ford horse; I'll get to that later). The harder he pulled, the more the horse would pull back. As he tugged on the reins, a voice-over narrator says, "Remember when you just couldn't get your horse started some mornings?" In a future installment, I'll go into gory detail just how hard it was to get that horse to cooperate or get started. It is also a good practice to time all the various shots in your spot. Hopefully, the reason for this should be obvious. The time for the opening "horse shot" was five seconds.

The horse shot would then dissolve to a woman dressed in 1920s garb trying to crank start her Model T Ford. As she bends down, dressed in her finest touring clothes, she tries to turn over the engine with a hand crank. As she pulls against the crank, she falls and slides underneath the car. The voice-over narrator says, "Remember when starting that old clunker meant major surgery." The running time for this shot was also five seconds. That left a grand total of 19.5 seconds to tell the who, when, why, where, and how much. The opening 10 seconds of the spot were humorous, but I wrote them in to establish the mood of the piece. I wanted to show all the things people had to go through in the "olden days" with their favorite modes of transportation. It would have been so much easier if they just bought a car at the Ford dealership. But using these ludicrous examples, hopefully I would get people to think, laugh, and wonder just where this idea would be going. Because I wasn't going to be airing the spot during the Super Bowl, I had a smaller budget and the viewer might not realize immediately that they were watching a local spot.

Most people don't have horses or Model Ts lying around. This is where homework comes into play. Visit local riding stables and tell them you are shooting a commercial (low budget). Every place I ever visited always let me use the horse for free if one of their employees was on the set (liability reasons), the animal would not be mistreated in any way, and they got a copy of the spot when it was completed. This approach works with "borrowing" almost any animal. I did have to borrow a camel once, and that cost a few dollars.

The Model T was more difficult to obtain. Check antique car clubs and ask around who might own the particular vehicle you need. Again, these people are very proud of their restored possession and *love* the world to see it. Promise them a copy and allow them to drive the car to the set, and no money will exchange hands.

Use your power in video for good. If you need anything for a no-budget video shoot, ask around to see if you can get it for no cost. As long as you don't abuse the privilege and give the owner a copy of the tape, you should get it for nothing. In almost 1,400 videos I've shot to date, less than 10 percent of the props had to be paid for (and that's because they were exotic and hard to find).

The World's a Car Commercial

I decided to forgo the sepia-tinted look or take all the chroma out and shoot in black and white; instead I shot in color. Even though I was talking about period things (a horse and a Model T), I wanted the spot to look modern (adding to the humor) and to have that lifelike taped look.

I also had written a third humorous shot to be used in place of one of the other two shots. I had envisioned a man pulling the Model T up to a modern AM/PM mini-mart and stopping in front of the self-serve gas pumps. He gets out of the car, nozzle in hand, and tries to figure out where to put the gas in. The voice-over narrator would say, "Having a hard time finding gas for your car?" When we tried to shoot the scene, it just didn't look as good as it originally did on paper, so it was cut. It was then my client demanded that I spend the rest of the time in the spot talking about the current great deals they were offering. The voice-over narrator would segue from the horse and Model T by saying, "Isn't it time to get a new or used car or truck at Crazy Fred's Ford?" While the narrator talked, we'd be showing visuals of the dealership. His five-second voice-over made the transition into the donut part of the spot.

The last 14.5 seconds would change every month with the current specials they would be offering. In the remaining time, two used cars and one new car would be featured. If they happened to sell the used cars before the spot aired, the viewer would have to read the fine print at the bottom of the ad that clearly stated that these particular models might not be in stock. Your butt must always be covered when writing a car commercial.

Hopefully, when it's all said and done, people will talk about the commercial and remember probably not more than one scene. I can only hope it was one of the first two scenes and not a particular car at the end of the spot. But if they walk away and tell someone, "Hey, did you see that commercial where that lady slides under that old car?" "Oh, yeah! That's the one where that guy can't get his horse started in the morning!" I rest my case.

Review: The Ten Most Important Things to Remember When Writing a Commercial about Cars

1. Use the product name in the spot as much as possible. The client will like you more when you do this.
2. Determine if the spot is going to be hard sell or soft sell. The old Ford spots used to be hard sell; I was doing the client a favor and made it soft sell. Once you decide on a sell type, try not to switch back and forth.
3. Try adding a touch of humor in writing your spots. Notice I said "a touch." No silly funeral home commercials. The viewer will remember a humorous spot longer than a serious spot.
4. Determine the type of audio you want to use: voice over, someone on camera speaking, a talking horse, and so on.
5. The humor or pace of editing should build throughout the spot. Let that last humorous bit leave a lasting impression (especially when writing a cement commercial).
6. If the visuals can stand by themselves (can you say it without words), don't use words to bog you down. I personally think both the horse and Model T shots needed a little extra audio plug.
7. Write more scenes or shots than you have time for in the length of the spot. Sometimes things look great on paper, but don't work out when shot.
8. Try to mention only the positive things about the product. This is especially true with car leases. Only the cheap monthly price is mentioned in big letters; all of the disclaimers and restrictions are in fine print.
9. Don't tell the viewer too much; always leave them wanting more. If possible, try to end the spot on a humorous note as well. Everyone likes closure (especially girlfriends).
10. Try to come up with as many ways as possible of saying the same thing without being redundant. The client is bound to like one of them.

The most important thing I learned from this Ford shoot was to not get discouraged when my best ideas got rejected by the client. I actually came up with eight different horse versus Model T opens before the client picked two he liked. The other ones will probably never see the light of day. What does the client know anyway?

Shooting on Location: Expecting the Worst

One of the best things about being in the television business is expecting the unexpected. That is just what happened on one of our recent shoots. When on location, you never have a clue what to expect. You will be fighting against the elements (heat or cold), your equipment (what happens if something malfunctions), the location (you are on the client's turf), and the public (if the business is open to the public, you are at their mercy).

Amused at an Amusement Park

My client was the marketing/public relations director of an amusement park. This park was built especially for kids but was more like a mini-Disney World. All the multimillion dollar rides and attractions appealed to adults as well as children. In this type of park, you could pay a per-person fee and ride as many of the rides as you'd like.

Our initial meeting with the client wasn't what I had expected. After talking to the 11-year-old manning the ticket gate and convincing him that I really worked in video, I was sent into the main office building. There I met a much older woman who was answering phones and connecting people with the parties they desired. What amazed me was that she could be talking on the phone, transferring calls, chewing gum, and filing her nails all at the same time. I also appreciated the way the designers of the park elevated her desk nine feet in the air so you had to look up to talk to her. When I told her I wanted to see the marketing person, she stepped down from her crystal throne and led the way through a maze of doors. Although she couldn't have been more than 17, she knew exactly where she was going in life.

The client was seated behind an immense mahogany desk that stretched out to each corner of the room. You can tell a lot about a client from their surroundings. I knew I was in trouble already (but don't let that jade you). Although on the phone, she greeted me with the usual wave of her hand and had me sit on a child-sized seat that was placed by her desk. When I had her full attention, she said that she wanted to videotape all the rides and feature them in a 30-second commercial. Besides it being the fastest moving spot in history with one second of screen time per ride, I offered a few suggestions of my own that might make the spot slightly slower paced and less expensive.

I told her that the best way to show most of the rides was to do it subjectively and that we would rent a lipstick camera to get *point of view* (POV) shots of her major ride attractions. People enjoy POV shots from the front seats of roller coasters, water flumes, and so on. She said to do what I thought best as long as it would look like it was fun for the viewer.

A Really Big Shoe

When the day of the shoot arrived, we were ready for anything. We had rented an Elmo MN-401X lipstick camera with a 4mm wide-angle lens that gave an excellent panoramic view. A lipstick camera was given its name because it is the actual size of a tube of lipstick. The camera was connected via a 30-foot extender line to a *central control unit* (CCU) unit that was powered by a BP-90 battery (battery #1). The CCU unit was attached to our portable Beta SP deck with a one-foot BNC line. The portable deck was also powered with a BP-90 (battery #2). In order to view what the camera saw, we had to also attach a portable three-inch color monitor to our deck; this was also powered strangely enough with a BP-90 (battery #3). The CCU, portable Beta deck, and monitor were all strapped together with bungee cords and gaffer tape. The tiny, 3-ounce lipstick camera with a lens had to be attached to this 40-pound recording unit. We must have looked like the bomb squad when we walked into the park with this contraption strapped to our chests.

We found out that having a monitor is crucial when you have a camera this small. It's very easy for the tiny camera to be tilted without the camera operator knowing it. Unless you want a Dutch tilt on every shot, the monitor must be constantly checked and the lipstick camera persistently adjusted. It was fun seeing what an ant's view of the world was like.

Our first subjective shot was going to be that of a merry-go-round. Although it doesn't appear to be a difficult or exciting ride, it proved to be just the opposite. We gaffer-taped the lipstick camera to the top of one of the horses' heads and rested the 40-pound pack on one of the adult seats behind the horse. Since the park was open for business and we were dealing with hundreds of overenergetic, sugar-fueled kids on every ride, we had very little time to set up.

When we were all set up (we had ridden the ride at least three times to find the best spot and best horse for the camera), we started our fourth ride on the carousel. The intense heat of the sun and the hot plastic surface of the horse caused our camera to shift slightly. Although the camera was durable, the slightest touch to the lens area adjusted the focus and tilt. It looked as if we were shooting a drug-induced, hallucinogenic trip through a carousel ride in the '60s.

I powered up all of our units and retied all the bungees. As the horse moved up and down and the carousel spun clockwise, I stared into the monitor to make sure that the camera didn't slip again. It didn't, but I was starting to feel sick. Here I was, a grown man getting sick on a merry-go-round. The horses were all looking at me now and laughing. The endless calliope music was also distorted and mocking me. As the camera operator rode the iris (via our strapped CCU) to keep the shot from darkening every time we passed the sun, I tried to think of nonsickening thoughts. At this point, it was kiddy park 1, director 0. The lesson learned here: Don't focus on the monitor too long. If you're body is moving and you are watching a stationary object, your equilibrium is thrown off.

The next POV ride proved to be less uncomfortable: the wooden roller coaster. As a child, I could ride a roller coaster for hours and anticipate each hill. That had suddenly changed as I had become an adult. Once again, because the park was in full swing, they didn't want to stop the roller coaster and let us attach our camera properly. When the car came screeching into the station, everyone jumped out and the camera operator and I piled into the front seat. In the 15 seconds we had before the car left and began to climb its first hill, we gaffer-taped the lipstick camera to the front of the car. We would get an unobstructed view of the entire ride.

As the 15-year-old, prepubescent operator squeaked "all aboard," the car rocketed to life. After an initial whiplash when the car first attaches itself to the chain that pulls it up the hill, we were ready for anything. The look of terror on the other passengers' faces should have been my first clue that I'd be in for the ride of my life. When we reached the top of the first hill, all I could remember saying to myself was AAAAAAAAHHHHHHHHHHH-HHH! I tried to look at the monitor to make sure that the camera wasn't tilting, but every time I glanced down the car would whip and bank around a turn, throwing me into my camera operator. His baseball cap had blown off on the first hill, my glasses were steaming up from the humidity of the day, and we were sliding around in our own sweat on the cheap vinyl seats. There was no way I could make anything out in the monitor. The screaming (in addition to our own) was deafening, as the other passengers in the car held their hands in the air. We, on the other hand, were holding on to our equipment for dear life (the video stuff too!).

I thought I saw my life pass before my eyes in the monitor, but I couldn't be sure because it was vibrating so violently. In less than 40 seconds, the ride was over and we were back at the station.

Since I couldn't see anything on the monitor, I suggested that we try the ride again with the camera turned toward us, instead of out towards the tracks and the colossal hills. This way I could get the perspective of the ride and still see the expressions on the people's faces (and record my own death). In our 12 seconds that remained, I turned the camera around and leveled it as best as I could and gaffer-taped it down.

When we finished our second ride through Hell, the camera operator suggested that we try the ride again, this time from the last car. We could now get the POV of everyone in the car and still see what was ahead of them. He ripped the camera from the front car and I grabbed our pack of power. It seems that all of our sweat and the blazing heat had loosened the tape and bungee cords that held our recording world together. As soon as I picked it up, everything came apart and crashed to the bottom of the car. Batteries came disconnected and the monitor slid out of its Pro-pack. I scrambled to pick everything up as the camera operator raced to the back car. Like in the cartoons, his leash was only so long, and soon it snapped him to a halt. Scooter, operating the ride, thought that this Laurel & Hardy routine with two grown-ups was humorous.

Not willing to give us a break, Scooter started the ride the split second our butts hit the last car seat. The camera was again attached to the seat in front of us, and I tried to connect the batteries to all their respective units. Luckily, they all were the same sizes. Needless to say, we had to ride the roller coaster three more times from the back seat before I had everything connected and strapped together. I felt like I was in a foxhole in the middle of a war trying to reassemble my rifle while being shot at.

Since I was too busy assembling equipment, the camera operator was able to watch the monitor. He said that the car was shaking too much to get a usable, steady picture. Shock absorbers aren't an option on roller coaster cars. He suggested that he tape the lipstick camera to the side of his glasses and try it from that vantage point. His body would act as a shock absorber, something the roller coaster car wasn't doing. If I didn't have hemorrhoids after this ride, I never would. I guess that's why kid's butts are so small; they can ride the roller coaster forever.

Five more rides in the back seat with our bodies acting as shock absorbers and I was ready to die. Besides being soaking wet from sweat, high-strung from a bad set of roller coaster nerves, and being pointed and laughed at by little moppets a fourth our age, I was ready to volunteer for active combat duty.

The client, helping us to our feet and mysteriously absent from all the previously mentioned fun, promised that the next ride would be tamer.

After we regained our land legs and reassembled the camera package, plus a change of three batteries (it's just like changing an infant's diapers), we viewed what we had shot. The camera attached to the glasses was our best bet. Although extremely fast, the footage was usable. We were off to the next ride. Lesson learned here: Shoot 15 times the amount of footage you think you'll need. Tape is inexpensive and getting a new set of nerves adds far more to the budget.

Our next ride was one of those rides that had 50 little swings suspended from a tower. When the ride started, the passenger would be lifted 25 feet into the air and spun counterclockwise. At that point, centrifugal force would make each swing extend outward, so that the passenger could swing horizontally.

Since it wouldn't be safe enough for the camera operator and me to be tethered between two different seats, I volunteered to remain on the ground and direct from there (membership has its privileges; I probably wasn't as tall as Bugs Bunny's ears at that point anyway).

The camera operator attached the camera to the swing's metal chain (the swing wouldn't shake at this point because it would be flying freely through the air.) I gaffer-taped the whole recording and viewing unit to his T-shirt and he entered his individual swing. If he didn't look like a terrorist with a bomb strapped to his chest, no one did. As he tuned out the sounds of kids yelling "Hey, Mommy, look at that funny man," the ride began. Looking at the monitor strapped to his chest, he was hoisted up and began to swing around and around and around. If the merry-go-round had gotten me sick, this ride could do in a gorilla. For all I know, directing from the ground, he could have been watching old *I Love Lucy* reruns on the monitor. He was easily the poster child for "what's wrong with this picture." Lesson learned here: If there's no room for two people, be the smarter one and stay on the ground. It could cost more in insurance benefits if both of you died.

On the next ride, being a monkey or having a tail would have been an asset. It was called the "Starcruiser" because it was shaped like a huge boat and was connected at the top with two steel beams. The entire ship would swing back and forth as it gained altitude. When the boat reached the nine o'clock position, it would slowly fall backwards and swing to the three o'clock position. This would happen over and over; each time the ship would gain altitude.

At least this ride's operator, Muffy, stopped the ride long enough for me to assemble the camera setup. The kids on the ride giggled, and their parents gasped as I climbed up to the top of the 15-foot-tall tail fin to gaffer-tape the camera to its metal surface. From that vantage point, I could see the entire spaceship, all its occupants, and several birds trying to nest in the trees.

As luck would have it, 20 feet below in the spaceship's seat, the camera operator said he had no picture. After wiggling every cable, replacing every battery, and chanting a magic spell, we found out that the gaffer tape had inadvertently turned off one of the eight switches that needed to be on (on the CCU).

As I rappelled down the surface of the tail fin to the cheers of the riders, I originally thought that they were happy because I had solved the camera problem (the whole park was aware of it by this time). But in reality, they were happy that the ride would now begin, and I wouldn't be holding them up anymore.

Moments after the camera operator and I sat in the back seat and looked at the monitor, the spaceship swung back and forth. At this point, I learned that it is best to block out the horizon in order to avoid getting seasick. Two kids in front of me got sick (which we caught on video) and they weren't even looking at the monitor. We would only be using a few seconds of footage, but we had to enjoy the ride several times to make sure we had enough usable video. After swinging in a spaceship for six rides, it's hard to climb a tail fin safely to retrieve a rented camera. Lesson learned here: When equipment malfunctions (and it will), look for the most obvious first, such as the On/Off switch.

Our next ride turned out to be rather disappointing. Sitting on a piece of reinforced burlap and sliding down a giant sliding board on your rear end is more enjoyable to children. We hooked two burlap pads together and the camera operator and I went down, hoping to catch part of my legs in the shot (this would have been a great POV shot). But as luck would have it, something sticky was baking on the slide's surface in the hot sun (some kid must have dropped a Slushy). Since I was the first down the slide, I guess the accident was my fault. The authorities are now calling it "the Pile-up on the Giant Slide."

I was descending at a reasonable clip when I hit the nonslippery spot. Immediately my burlap came to a complete halt, but I didn't. Maybe it was the 40-pound pack that was strapped to me. I went tumbling into the next lane, tethered to my assistant. The CCU line was now strung across most of the width of the giant slide; batteries were lying everywhere. Soon the kids started piling up. You would think that when they saw the line stretched across the slide that they would try to avoid it. Oh, no. They headed directly toward it, laughing maniacally. After about 12 of us were in a clump in the middle of the slide, the first bunch of kids asked us if we could do it again. I politely said no. I've never been banned from a kiddy ride before. Lesson learned here: When moving apart from your tethered compatriot, unhook your BNC line. This causes fewer problems and means more in the lawsuit.

In the hot sun, at least the next ride would prove to be refreshing: the water flume. With four people per car, we would ascend an open vinyl tube and then float through the ride's water-filled trough. We got in the back area of one of the plastic, tree-trunk-carved, Flintstone-like boats and attached our lipstick camera to the back of the front seat. The water would easily cushion the ride. We sat back and relaxed; this would be easy. After all of our struggling, this ride would be a piece of cake.

A small child of about six asked if he could ride in the front of our car. He was short so I thought he wouldn't obstruct our view. I said okay. I still shudder when I hear the next words he spoke: "Come on, Dad!" Wouldn't you know, his dad would be a 600-pound linebacker! I've never seen a human so large. He barely fit into the car. He had more hair on his back than most bears. As his great mass settled into the car, we displaced half of the water in the ride. This behemoth was totally blocking all of our view and most of the sun. He was a human eclipse. We could have used his white T-shirt to screen Panavision movies. It was too late, however; the ride had started.

We removed the camera from the seat and hoped for the best. I've gotten less wet taking a shower. Huge walls of water would cascade into our boat because of Goliath's massive girth. Each time we adults in the back got soaked, little Goliath laughed hysterically. We covered all the video equipment with our saturated bodies, and in the hot sun we soon began to smell like wet dogs.

We were finally able to get the shot because the next time some kid asked if he could ride in the boat with us, I said that we were contagious. Lesson learned here: When someone asks if they can sit in front of you when taping, just say no unless the script calls for it.

I made a VHS window burn for the client in order for her to choose the best portions of each take. We now had enough footage to make a feature film.

The best way to sum up an experience like this is difficult. It's like trying to prepare for a trip to the tropics. We had the smallest camera known to man, but we also had to lug 40 pounds of deck, batteries, and monitor. I wouldn't have wanted to attempt a shoot like this with a conventional Beta-cam camera.

If anyone would ever like to know what a POV of a particular ride looks like, I can tell them from memory. I have a call in to the folks at the *Guinness Book of World Records*.

The reason I related this lengthy story is that I was never more unprepared for a shoot in my life. I had the initial meeting and knew what we were to shoot, but I didn't know all the problems we would encounter. You are going to be up against elements you have no control over. I now have an aversion to kiddy amusement parks. You now know what not to do; try to learn from my mistakes.

Review: How to Better Expect the Unexpected

Although these "tips" are amusement park related, substitute the word "ride" with what you are shooting.

1. On your initial visit to the facility, find out what you're going to shoot. Be prepared to spend the day at the park and ride the rides that will be featured in the spot. This is the best way to become familiar with them and where best to put the camera. Look at each ride from the viewer's perspective, what would they want to see? Subjective works best in roller coasters and viewing people enjoying a ride is better in a merry-go-round (leave it alone; I'll never mention that kind of amusement ride again!).

2. Establish every ride first in a long shot so people know what you are talking about. Immediately after that shot, go subjective. This is where a storyboard is helpful because each ride needs to be shot from a different perspective and the day of the shoot is no time to be fumbling around trying to plan this out (of course, that's exactly how I did it).

3. Take control in the first meeting with the client. Listen to what they want and then tell them how you'd like to show that happening. You are at an amusement park, so it should be a fun experience. Even if you have a deadly fear of clowns, this is not a circus and the clowns aren't real (anymore).

4. Keep the equipment to a minimum. Backup gear and batteries are important, but take the smallest, lightest equipment you have and leave yourself three times the amount of time you need. You're fighting against many elements and it's difficult to plan for every contingency (like when you get sick).

5. Make sure you have a monitor with you at all times to check what you are seeing. Playback is also critical because the next day is too late to find out the shot you recorded isn't there (just like the spinning jackals you saw after riding the merry-go-round).

6. Use gaffer tape, bungees, twine, rope, bandages, staples, or any other strapping device to hold things down (like your lunch on the merry . . . sorry). You will find these adhering devices are extremely useful in ways you never thought possible. Just make sure you have enough (remember to share).

7. The amusement park is open for business and you have to work around the paying customers (unless they made you pay also). Their entertainment comes first (not by watching you screw up) and you have to keep that in mind. This will

(continued)

(continued)

slow you down (at the least), but be patient (like when I was a doctor before I had any patience).

8. Check the ride on the day of the shoot a few times without shooting. The sun could be in a different position and your storyboarded shot might not work. This isn't an excuse to ride the rides free more often (that's what I told the other kids).

9. Check your equipment constantly. Movement, heat, cold, dark of night, and sun will change often so make sure you are shooting with something that's functioning. It's better to overcheck than undercheck (but never rubber check).

10. Motion sickness is a definite possibility because you are getting older and you're doing it more often than most. If you think you may become ill, prevention (before the riding) is the only way to avoid it. Take legal drugs and if that isn't possible, try to focus on something else to take your mind of the sickness (like a slowly moving ride with horses).

11. Don't focus all your attention on the monitor or you will become sick. Just like avoiding eyestrain at a computer or highway hypnosis when driving, glance at different things constantly. This allows you to see the horizon and know that you are actually moving.

12. The same thing applies when shooting. Keep both eyes open (the one in the viewfinder and the other) elsewhere so you can perceive what's happening around you. I got air sick (it seems like I always am) while shooting from a plane using only one eye (I'm not a Cyclops). Having no drugs (it wasn't that kind of flight), I was told by the pilot to fly the plane myself and that took my mind off my sickness (and made it focus on how I was going to kill us all because I never flew a plane before).

13. Don't change the angle too often within the same ride. You probably can't use 28 different angles of the same ride in a 30-second spot. Beforehand, decide which angle works best and limit it to no more than three, which still leaves you many options in editing (like why you got into this mess in the first place).

14. Use as few wires as possible if tethered to another person. These cords will cause tripping, falling, tangling, and suing at the worst times. This is where wireless is best ("Look, Mom, no wires!").

15. Take frequent breaks between rides to allow your stomach to catch up. This is a good time to play back the footage before you leave the particular ride. Do this discreetly because everyone will want to see it too (they just rode the ride; why would they want to see it again?).

16. Have too many spare batteries, drink plenty of fluids, wear lots of sunscreen, take two aspirins, and go to bed (alone). You may not realize that you are dehydrating trying to capture that elusive shot. Have spare and backups of everything. You can't buy a fresh BP-90 at the park.

17. Shoot miles of footage of everything because not all of it will be usable because of shake, jitter, focus, exposure, and color balance. These items should be checked before the ride, but you won't have control over most of the ride (believe me, I lost control long ago).

18. If the motion of a particular ride makes you sick (like plastic horses) have someone else shoot that shot. You can try to be a hero, but most likely you will be a vomiting hero.

19. Never stop the ride so you can adjust your equipment. Ride stoppage, like a movie abruptly ending after three minutes, does not win you friends, and you are far outnumbered by the "little people." Check your equipment before boarding and use two or three ride attempts to set up before shooting. This keeps the youngsters happier and they will whine less.

20. If any equipment malfunctions, check the most obvious things first. The On/Off switch is more apt to change than blowing a video head in the middle of a take. Take your time, don't panic, and you will discover the problem before having to ask the occupant in the next seat to help you.

The Corporate Video

Creating Nonboring Corporate Videos

Most, including myself, always thought that corporate videos were dull, boring videos having only slightly more appeal than root canal. But after producing over 400 in my sorted career, I'm now singing a slightly different tune. Actually, having more than 30 seconds to tell your story opens new avenues once thought nonexistent.

In this chapter, we'll discuss ways to break the mold around boring corporate videos. People who work in businesses (and want videos) have a life and want to be entertained when watching a video. I'm sure you remember watching some dull corporate videos back in high school. The producer thought exposing the young mind to corporate America through film or video would enlighten the viewer. Often it did just the opposite.

Let's start with a definition of what a corporate video is. My students in Video II are asked to do a corporate video as one of their projects. Since they always ask me what the term means, I'll explain it here also. A corporate video is a program that educates the viewer on what a particular company does or what service it provides. It may be an informative piece on how "X Company" helps the environment, economy, or local businesses. I've often done videos that simply show what a business does and how they do it well. I'll explain in detail a specific corporate video shortly.

By using creativity, your corporate videos can be interesting, entertaining, and informative. Figure 4-1 is an example of smoking on the set.

Smoke Gets in Your Eyes

Sometimes doing things that are supposed to be good for you can make you ill. I recently had the opportunity to make a corporate video on ozone awareness. So now that I'm aware of it, why don't I feel so good?

The client wanted to make a video for kids on what ozone is, how it protects us, and what we can do to protect it. The video would also show how his company (who was paying for it) was doing their part in the community. The video would be chock full of simple suggestions on what to do and what not to do to help the environment.

The setup was simple. An actor playing a Bill Nye, Mr. Science-type character would explain to the viewers what ozone does (it protects us from becoming toast) and what actions we can take to prevent its depletion (don't

Figure 4-1
On the set
with smoke

drive, cut the grass, or set fire to anyone). I know this all sounds like another boring corporate video, but we had some fun with it.

The first thing the actor did was tell us what ozone protects us from becoming. He said tongue in cheek that the ozone protects us from harmful ultraviolet rays. If the ozone wasn't there, our skin would look and feel like an alligator's (he never said anything about the teeth though).

In order to stress the point, he produced a baby alligator and plopped it down on the table in front of him. The foot-and-a-half-long star must not have read his lines because he immediately wanted no part of us. The creature would have to be held to be contained on the set. We had a teleprompter set up for the actor (and the alligator if he so chose) with our camera directly behind. As the actor began reading his lines in preparation for the gag, the alligator began squeaking. I don't know if he was a quart low, wanted to be back in the water, or wasn't happy with the script. This would be the first scene we had to audio dub. There's nothing like going through the trouble of locating a real alligator only to have it sound like a pet's squeak toy. Like I mentioned in the last chapter, this "borrowed" creature cost us nothing because we asked, said we had no money, and gave the pet shop guy a copy of the tape.

The next scene involved fog. The actor was to walk into a smoke-filled area (simulating smog), cough, and discuss the harmful effects of smog. That obviously doesn't sound like anything that's too difficult to shoot. The actor, shrouded in smog, was to talk to his sidekick (a little girl) about the effects of this "smoke." She would be standing 10 feet in front of him, so the smoke wasn't supposed to get near her at all. She would remain in the clear, and he would be wheezing his lungs out.

The scene was shot in a 30- by 20-foot enclosed studio. The lights were suspended on a grid 20 feet above the floor. The air conditioning was on to keep us from baking in our own juices. We borrowed a fog machine from a local rental house (we asked to borrow it to make sure it worked in time for Halloween). It's nice to know that anyone who wants to rent a shovel, weedwacker, or tent can get a fog machine from the same place.

If you've ever operated a fog machine before, you know they all operate in the same manner. The unit is plugged in and allowed to heat up. Fog juice is released onto a hot surface and chemical fog is created. A timer can control this fog so a measured amount can be released at a given interval or you can manually have fog on command. We opted for the latter.

We set the adult actor five feet from the cyc and had the girl in front of him. The fog machine was stationed eight feet from the actor, camera left. A piece of foam core was laid over the top of the fogger to keep the smoke from rising too fast and getting circulated through the studio.

On the first few rehearsals, I was the button man. When the director called action, I pushed the button and the machine spewed fog. The fog would encircle the actor, obscuring him and the little girl. Time for plan two: Enter the wafter.

Because I needed only one finger (just like George Jetson) to operate the fogger, I had other appendages that had nothing to do. I was therefore elected to wave a four-foot piece of foam core violently behind the girl to keep the fog from enveloping her. It's not as easy as it looks. Here I was flapping away like some gooney bird in heat just to keep the fog from wafting in front of her. It didn't work. The harder and faster I wafted, the more wind I created. On the upstroke, I pulled the fog toward me, and on the down stroke the wind I created made her hair blow. I didn't look convincing. A four-foot piece of foam core gets mighty heavy when you're flapping it all day.

I was now given instructions from every member of the cast and crew on the correct way to waft. I had no idea that it was such an acquired talent. After 20 takes, I finally learned the wafting art. It was time for a break. We opened the doors of the studio with fog emerging behind us. It looked as if we had just left a screening of a Cheech and Chong movie. Luckily, the fog was nontoxic or I'd be a wafter in heaven.

Fog is supposed to stay low to the floor, like dry ice in water does. Dry ice produces a bubbling effect not quite the style we were looking for.

Our next plan was to use a smoke cookie. The fog worked pretty well covering the adult actor; we just needed it to envelop him on cue. We decided to use the fogger behind him and use a smoke cookie in front of the camera.

We had purchased a smoke cookie (about $2) that when ignited would produce white smoke. Although resembling a sugar cookie, it would be lethal in my hands. Only a small chunk would be needed for the effect. George Winchell, our smoke technician, broke off a small chunk of the cookie and placed it in an empty 35mm metal can (100 ft null size) lid. A black foil wrap tent was placed over the cookie and film lid. A hole was punched in the center of the foil tent to allow a plume of smoke to billow up and in front of the teleprompter hood. There it would encircle the lens. At the exact right moment I would light the edge of the cookie with a cigarette lighter. Since I had been removed from my wafting position, I needed something else to do.

On cue, I would light the cookie and smoke would blend in with the fog to totally cover the actor. The fog would flow behind him and the real smoke would flow safely in front of him. Now I know why my mother never let me play with matches as a child. If I smoked using a lighter, I know I'd set my beard on fire.

At precisely the right moment, I carefully lit the edge of the smoke cookie. It immediately started sparking and smoking (I expected the smoke but not the sparks). Soon yellow flames were pouring over the top of our tent and thick white smoke covered everything. It worked, but the actor was obscured with smoke too quickly. I would have to ignite another chunk at a later line in the script.

So, take after take and time after time, I kept lighting the cookies. They kept sparking, smoking, and burning. We still couldn't get the timing right. The little girl and her family had long since left the studio for this effect. The smoke cookie produces an unpleasant but totally nontoxic odor. I was now feeling like an overcooked spare rib.

Finally, on the twenty-sixth take, we had our timing down to perfection. I knew the exact word in the script to ignite the cookie on, we had the perfect number of holes in the foil tent, and I had learned to use the lighter without setting my finger on fire. I also learned that the metal on the top of a lighter gets extremely hot when you have the lighter burning for 20 seconds 30 times a day.

As soon as the director called action, the smoke alarm in the studio sounded. I was standing next to the smoke cookie with a burning lighter in my hand. What more proof did the fire department need?

All the fog had created no noise from the smoke alarm, but the cookie smoke was a different story. It seems as if the air conditioning in the studio had picked up the smoke and was circulating it through the duct work in the building. Within moments, the halls were filled with people running around wondering where the fire was. Because of the local fire code, the building had to be evacuated as the ear-piercing chirps of the fire alarm droned on. Since we had to leave the studio, I grabbed the nearest thing, the lighter, and left the studio.

Why I grabbed this lighter I'll never know, but here I was walking down the crowded hallway with a lighter in my hand. I was now an arsonist or a really bad chain smoker.

I had no future as a wafter, but I could make something out of my life as a cookie igniter. The conversation outside the building was basically, "Who set off the fire alarm?" Of course, it was me, but the director had told me to do it.

With my career in video smoke now over, I could be rented out for children's parties if you supply the lighters. The rest of the video, what was left of it, went off without a hitch. The fifth-grade viewing audience enjoyed the video and laughed at the appropriate places.

In order to make your corporate video entertaining, you should add to the script what it needs to make it enjoyable to you. If you don't fall asleep writing it, the viewer might also stay awake. I shot a video at a Twizzler's factory (the licorice company). The script wasn't that thought provoking, but the images of sweet, red, luscious-looking licorice made everyone enjoy the program. Here the visuals were the enticing factor (or flavor).

I made another video about a machine that inspects closures (bottle caps). How do you make something repetitive and visually uninteresting not so? The answer: Write it into the script. With an interesting soundtrack (music, effects, and narration), you can make a visually dull video more exciting.

The last bit on this subject is the narration. With an enthusiastic narrator, even casket making will keep the viewer glued to the screen. If the voice over is slow, dull, and monotone, you will lose the viewer immediately. The producers and creative people should be enthused about the project. Include the narrator and you're well on the way to an interesting corporate video.

Review: Six Steps to Keep Your Corporate Videos Exciting

1. When you need animals in your corporate production, ask pet stores or owners if you can borrow them for the afternoon. Keep the amount of time they are away at a minimum. If you have them too long, you may end up owning that paranoid ferret.

2. It takes 12 times as long as you think it will when working with any animal. We bipeds are usually the smarter objects and we can take direction. If you obtain a creature from a professional wrangler, they will take stage direction (the animal and the wrangler), but you will end up paying for that luxury. Instead just use the muppet as little as possible to get the point across to the viewer. What do you think chickens do before they end up in soup?

3. When using special effects like fog, you even have less control. Fog will lie there, hover, or disappear when you don't want it to. It's hard to get just the right amount; you always have too much or too little. Wafting is the only way to "place" fog where it needs to be. It takes practice to position the fog, but it does move easily. You just have to know how to control it. Normally, it will cling to the ground unless air causes it to move or rise. You should be that air. Just don't be full of hot air.

4. If you need thicker fog, use smoke cookies. This smoke will billow and fill the space quickly. Black wrap foil is the best way to enclose the plume so the smoke has some direction. Be careful igniting these cookies and try not to breathe. Leave enough time for the smoke to dissipate or hire as many chain smokers as you can find and have them all take a deep breath at once.

5. Make the dull interesting by writing an exciting script. Boring material can come to life if you punch it up in the scripting process. Try adding music; this can turn the whole feel of a video around. Use magic words (adjectives) and you will cast a spell on your audience (unless they're witches already).

6. When using a narrator, have him or her read in an upbeat manner. Like the music, a narrator sets the pace for your project. If it's a touchy subject, have him or her read it accordingly. If they can't read it with interest, your audience will soon leave the room. Just make sure they pay on the way out.

Making the Most of What You Have to Work With

When you haven't been dealt much by the client, it's up to you to do the best you can with what you have. Of course, this means your budget is squat, the client has no clue how he or she wants the video done, and it's all up to you because "you're the professional." I really don't enjoy it when a client hires me because "you can make it look good" when they have no idea what they want.

Through another one of my amazing (but true) stories, I'll tell you how editing saved the day in a corporate video when the client didn't know what she wanted (and when I told her what she wanted she still didn't believe it).

Easy Editing Made Difficult

Sometimes you're handed a project that seems almost impossible from the start. The client wants it yesterday, you have miles of footage, and he, she, or it wants total approval of each process, but they have no idea what they want. Of course, totally understanding my many constraints, I accepted the project and enjoyed every second of it.

The client wanted 12 single-person interviews (they were married, not single, but wanted one person in each shot) taped in 12 different cities (from Maine to California) in the course of one week. Because my time machine was in the shop, I had to hire crews in each of these locations to perform the task of shooting. Of my newly assembled dirty dozen, I wasn't able to get every crew to shoot in the same format. I had crews shooting in Betacam SP, Betacam SX, DVCAM, and DVC-PRO.

Our main objective was to have all the footage look somewhat similar so it could be easily edited together. If the crew in Thumbtack, Wyoming shot it on Super 8 film, it could differ from that of the other crews. You can perform a few steps to keep things looking the same (without cloning). I created a detailed list of what each crew should do so the end result would look like the same crew had traveled the globe.

This was another area where lack of budget came into play. If you are asked to travel all over the globe for a video production, realize that travel costs quickly jack up a budget. You have time involved whether you're sitting in an airport or on a plane, hotel room expenses, car rentals, meals, and the logistics of everything. If others in the different locals are hired, your travel costs are eliminated, lowering the bottom line.

The lighting should be similar whether high or low key; the shot should be framed with the victim looking in the same basic direction, and any other feature you desire should be consistent so the renegade crews may follow your road map. I chose the *Dateline* or *20/20* style of interview because everyone seemed to be aware of that particular format. Getting Stone or Diane to travel on all these shoots wasn't easy either considering our budget.

Since we were receiving four different formats of tape, we had to transfer everything to the format of choice, ours being Betacam SP. Consistency is necessary in transferring, it doesn't have to be Beta, but choose a format you prefer. If you try to digitize a DVC-PRO tape directly into your editing system, it won't match the characteristics (good or bad) of Beta. Transfer the master tape to Beta first and then digitize it. Now everything is at least on the same page.

Once the many tapes came floating in, have every single word of the interviews transcribed for a paper edit. This is the only way to make sure the best pieces are used without wearing a hole in your tapes (or your wallet). If someone has twenty thousand "ahs" or "umms" in their first sentence, that could be noted on the transcription.

With 20 pages per location, use the nifty highlighting feature of your word processor and underscore the sound bites that make the most sense. I used to highlight each line with a yellow marker, but my mother thought I was jaundice and lacking something in my diet. I won't go into the truckload of wax beans I had to eat for eight months. Besides looking like canaries on paper, your highlighted transcribed script is the next step in the editing process.

Have the highlighted scripts reviewed by everyone who needs to see them. This way the client can't come back later and ask why someone didn't say something. Everything is in front of them in black and white (and some highlight color). This step also gives you a rough idea of the size of the rough edit. Our first cut looked like the movie *Intolerance* (that's how I also felt editing it).

When the best takes have been selected and approved, now it's time to enter everything in your nonlinear system. A cuts, A/B roll, or a conventional linear editing system will suffice, but changes to the edit will cause premature baldness. At this point, any *nonlinear editing* (NLE) system will work as long as you have the necessary storage space (I mean disk drives, not a closet). I was able to enter only three hours of the best footage with our tower of storage power. With a final running time of 25 minutes, I had a lot of cutting ahead of me.

Because of limited hard drive space (from other projects, not because I'm cheap and didn't buy enough), I stored all the footage at a lower quality to save disk real estate. Of course, every face was one giant pixel until I realized that's what most of the people looked like. Once the project is closer to its natural size, the system will batch digitize at a realistic resolution.

I won't go into the basics of nonlinear editing, but this type of creature enables you to rearrange the sound bites in their best order. At this stage, don't worry if nitwit one and nitwit nine don't edit together properly because of a jump cut. Just get everything in the order that best suits you.

Now that your rough cut stretches from here to eternity, it must be pared down to a more manageable size. How do you decide what is great stuff, not so great, or simply useless? This is where your years of experience, your keen eye, and maybe the client come in. Listen to the flow; are they repeating themselves in the same sound bite? Does the speaker scratch at an inopportune time? In most videos, the sound bits can be edited to remove the bad words and sounds because the visuals are B roll. In our case, the talking heads were the only visuals, eliminating the opportunity to edit for clarity.

The great thing about nonlinear is that you may edit and reedit as often as you like to intoxicate the viewer. Each time I would try to shorten the running time, I might only trim two minutes. By the time I was at version 621, I was pleased with the result and knew the Grand Poobah would also.

The key to editing a project like this is to change anything and everything until you're pleased with the results. Document each step, save everything, and keep your paper trail. Once the client got his grimy little hands on the rough cut, most of the shots were resequenced. Try that on an A/B roll system.

Review: Ten Great Things to Do When Using Multiple Crews and an NLE System

1. A great source for finding crews across the globe is www.mandy.com. We have used crews from all over the planet and never had a problem. This interview process is crucial to pick the right crew for the job. Just make sure you type "mandy" correctly. If you misspell it, the site it takes you to would upset your grandmother.

2. Don't pick a crew just by price alone. Check their demo reel, their distance from the location, and their business practices. One group wanted to be paid in small, unmarked bills (actually that's how the client is paying me), so we found someone else.

3. Make sure each crew is shooting in a similar style. The only way for that to happen is to spell out everything you want on tape in great detail. I told one crew I always wanted a pony as a child, and they sent me one! Too bad that kid's mother called the police on me. It seemed she wanted the pony back and not her kid.

4. Transfer all footage to a common tape format. Even if originated in another, you will have a little consistency if everything is digitized in a similar format. Our Fisher Price tape deck accepts only the highest quality eight-tracks.

5. Have every word transcribed from the interview tapes. If you do a paper edit, you will know exactly what was said, everyone who needs to be involved will have access to all the words, and you can easily cut and rearrange the sequences on paper before editing begins. I guess that's why it's called a paper edit instead of a paper cut; it hurts less.

6. Digitize the footage into your editing system and lay out in the order that works for your client. Don't worry about the length of your rough cut at this time. Sometimes words look great on paper, but the interviewee may have something hanging from their teeth in the video.

7. Rearrange the clips and delete footage until you are closer to your allotted running time. Make sure the person isn't saying the same things over and over. Pretend he or she is your nagging in-law. How much does he or she say that is really worthwhile? What can you cut out?

8. If your client is nit-picky, have them review each step of the process. It may take a little longer, but the client will live a longer, happier life. Overproducing is a way of life for some. Don't let it happen to you (brought to you by "The Producers of Video Foundation.")

9. After your 3,245th version of the edit, have a high- or normal-resolution screening for your client. Make sure everyone is happy before you start the duplication process. Now you may sing, "Send in the Clones."

10. Save the edit decision list (EDL) someplace where you can access it, since the client will come back and make changes again right after the last label has been attached to the final dub. Just remember, the client is always right, except when they're wrong, which is all the time.

Getting Light Where It Normally Isn't

If light was where you wanted it to be all the time, life would be wonderful. But times will occur when you need to illuminate some darkened crevice and seem to have no way of doing it. This section will discuss how to get light into places you never thought possible and how to do it inexpensively (cheap is too crass).

The Grannies Have It

Sometimes overexcited producers and directors want light emanating from a place where it wouldn't occur naturally. Forget about all you learned using practicals. Sometimes you *must* have that illumination where it really isn't practical. As an example, when was the last time you had to stick a light in a half-gallon carton of ice cream? I mean *not* when someone was mad at you. In any case, I don't think the producer was mad at me, but he wanted to have a light emanating from a container of ice cream sitting on the counter.

Picture, if you will, a kindly old granny walking into her kitchen and spying a container of ice cream sitting on the counter. If you know anything about grannies, mine included, they *love* ice cream. But, of course, this just wasn't any ordinary ice cream. This was a new flavor called Maximum. This was prescription ice cream. If someone went to the doctor and they needed ice cream to get well, this would be the ice cream that would fit the bill. Can you see the light that's hidden in Figure 4-2?

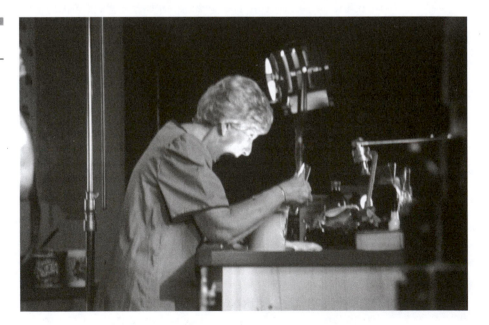

Figure 4-2
Light from ice cream

When Granny gets to the ice cream on the counter, she is supposed to open it. After looking around, she cautiously opens the lid and is blinded by this incredible blast of light. The director said that the light had to be strong enough to singe her eyebrows (I'm glad they cast a different granny than mine; singed facial hair wouldn't go over well at Thanksgiving).

So, I guess you would agree with me that the obvious way of getting a light to burst forth from a container of ice cream was to, everybody together, *put a light in the ice cream container*. Sounds easy enough, but even after I got the biggest grip I could find, he still couldn't shove that 2K into that half-gallon container. His excuse was that if he had actually gotten the light into the container, the plastic would have melted within seconds after the light was turned on.

Being a somewhat intelligent person, I suggested that we put a smaller light into the container, one small enough that we wouldn't be able to see it sticking five inches out of the top. The only light in our arsenal that would fit the bill was a pepper. Peppers are perfect little lights that don't take up too much space, pack a punch when it comes to output, and are tungsten balanced so you don't really have to color correct them.

I first tried a 250-watt pepper that we had in our lighting kit. I was going to use the 100-watt pepper, but if the director wanted a large amount of light, by golly I was going to give him that. Luckily, the ice cream container was plastic and not cardboard. As a precaution, I wrapped the inside of the container with black foil; this would at least keep the heat to a minimum. If you've used peppers before, you know that they really hold the heat. Long after the 5Ks and 2Ks have cooled off, the peppers are still too hot to touch. I was told this was because they were made of metal (so is my car for that matter and I can still touch it).

Anyway, I confidently dropped the pepper gently into the opening of the container; the unit still protruded two inches. Calmly, I removed the barn doors, glass lens, and any other miscellaneous hardware that added to its girth. Voilà, it fit. A small hole was cut in the rear of the container to allow the AC cord to be attached to an outlet.

In order to reflect even more light onto Granny's face, I installed a small piece of reflective foil to the underside of the lid. She was then instructed to open the lid toward herself, hiding the lighting unit, and the reflector would send a beam of light directly into her eyes. No, before the Granny Wranglers of America get up in arms, no grannies were mistreated or harmed in any way during this shoot.

Once the lid was closed, it was impossible to tell if the pepper was on or off. The black foil was really doing its job. However, because of the confined, enclosed space, the container would quickly heat up. So, because of my big mouth in pointing this out, my job for the next 23 takes was to switch on the pepper 1.32561 seconds before Granny opened the lid. I knew those 15 years in college weren't a waste of time.

So now everything was set. Granny and I were a well-oiled piece of machinery (I'm talking about our timing with the light and the lid; what do you think I meant?). She would walk into the kitchen with well-combed hair and wearing a nightgown, casually glance at the container of ice cream on the counter, look around, and then open the lid. The pepper's light would hit her face, she would shriek, and the camera would stop. If that were all that was involved, it would have been easy. But no, the director wanted more. It seems the blast of color-corrected tungsten light on her face wasn't quite enough. He had visions of fireworks, explosions, dancing bears, figure skaters swirling around Granny's head, and, oh yeah, smoke.

I sheepishly looked into my lighting kit and said I had used my last dancing bear and figure skater on a previous shoot. To my chagrin, he calmly said, "We'll add that in post." Now that he and I are on a first name basis again, if he asks me for something, I'll tell him we'll fix it post.

After all the effects, smoke, and lighting had done its thing, Granny closes the lid and says her only line, "Whew!" Her hair is now standing straight up, smoke is coming from her nostrils, and her canary, earlier calmly singing and swinging on his perch, was now medium well done and pushing up the daisies.

There you have it. Another crisis solved. But the director has that strange look in his eyes again. I hope it's only gas. I will never venture into the kitchen alone after dark looking for ice cream, not unless someone else opens the lid first.

How to Be a Bouncer

Now that you know a nifty trick for pointing light where it needs to be, what do you do if you can't get a light where you want it?

Not being built like a linebacker (I sometimes smell like one), the only kind of bouncer I'll ever be is someone who bounces light (there's less stress bouncing light anyway). Sit back and I'll tell you everything you wanted to know about bouncing but are still afraid to ask.

J. Josephson is a wall covering manufacturer in Hackensack, New Jersey. Large, 1-ton rolls of wall covering are loaded onto a machine and it cleverly spews out more manageable 30-yard rolls. In the past, several people had to operate such a machine that would create 30-yard rolls, but it was labor intensive and time consuming. This new machine can package enough wall covering in one day to paper a Las Vegas casino (about 750,000 square yards).

Our job was to produce an eight-minute corporate video on the operation of this new machine and give trade-show attendees a glimpse of the future. A CD-ROM would also be created for facility tours and meetings.

George Winchell directed the shoot and devised a lighting plan that was second to none. With the help of Tom Landis and myself, George bounced his way into the production.

George's initial lighting plan was to use 2,000-watt (2K), open-faced lights that would provide the key illumination for the 70-foot-long, 20-foot-high machine. Two lights were used at each end of the steel behemoth. Fill light on the wide establishing shot was supplied by three DPs. Each 1,000-watt (1K) DP was silked with 216 diffusion to soften the blow. The illumination in the plant was fluorescent lights with a little daylight spilling in through high windows. The 2Ks and 1Ks provided an even light that highlighted the correct parts of our star. Figure 4-3 shows how high you must get to shoot a machine of this size.

Figure 4-3
The best vantage point

Often wider, flatter, and stronger directional light is needed when shooting football-field-sized objects within a confined space. When it came time to shoot close-ups, a much softer approach was needed.

Our Magliner cart supported the camera case, deck, and 12-inch monitor. This rolling cart enabled us to roam around the machine at will as long as we were connected to an AC source.

The wide shots necessitated I climb aboard a Yale forklift (at least it was well educated) and be hoisted to a height of 17 feet. From this tiny platform, I had a view that few humans (and no mice) ever saw. As long as I was connected to the base unit via my 25-foot multipin cable, I could stay up there-all day (except for potty breaks).

After the long shots of the machine were captured on tape, it was time to show the various smaller aspects of the machine in close-up. If you've ever tried lighting a metal machine simply by blasting light into its deep crevices, you know it seldom works. The dark pits may be better illuminated, but the flat surfaces glow like an alcoholic after an all-night binge.

The angle at which the foam core is placed is important. A piece lying on the ground will provide fill light from underneath and an angled piece will send the light in a different direction. Take the time to experiment with the position of the bounce card; you can control exactly where the light falls.

With trusty foam core in hand, George mounted it to the side of our college-educated forklift. A DP was pointed at the foam core and the light bounced onto the green metal surfaces. The result was a pleasing, even illumination. Don't be afraid of using your tripod, forklift, or any other nearby object as a leaning post for your foam core. Sometimes C-stands are at a premium, so use what you have on hand.

On occasion, the foam core on my forklift platform would be turned on its side and used as a flag to block the key light. That's the great thing about foam core; it can bounce as well as block light. Before the foam core volunteered, my back acted as a light-blocking flag. If it weren't for the stench of my blistering and searing skin, I could have done that all day. You don't need fancy equipment to get light in hard to reach places, as shown in Figure 4-4.

Other times when we needed to shoot an extreme close-up of shiny gears, we would surround the area with foam core. Our key light would do its job of illuminating the area, and the foam core would bounce a softer light into the narrow field of view.

I asked George how he knew whether to use a 1,000-watt DP or 2,000-watt open-faced light with the foam core. Once he showed me the difference in lighting between the two, I gained new respect for the thin white board.

Figure 4-4
More bouncing

You will actually get less light on an object if you bounce an open-faced 2K into the foam core. However, using the focusing ability of a 1K DP, the spotted beam of the bounced light actually raises the illumination of the shot. This is one instance where less light bounced actually gives you more illumination.

The 2K might give you a broader or wider bounced source, but if you need more light in a given area, a light that can spot its beam will give you a higher light level (and therefore more respect).

The rest of the day we spent getting extreme close-ups within the bowels of the machine. In circumstances like these, direct light and bounced light can work hand in hand. The direct punch of a light is necessary when illuminating dark voids or cavities within the machine. In these cases, the barn doors on your light, black wrap, or even flags will keep light from spilling where it isn't needed. But for the rest of the area, the soft light from a bounced source keeps the shine and hot spots down to a minimum.

These are just two examples of inexpensive techniques for getting light on the subject. If you have the courage, read on and learn why I'm called frugal (not cheap).

Review: Six Ways to Use Lighting in a Pinch

1. Don't be afraid to put lights exactly where you want them: under items, inside containers, or behind objects. This is a much less expensive way of illuminating. Doing any effect like this in postproduction is expensive. Just remember to poke air holes so it can breathe.

2. Use smaller lights to get where big units can't. Tiny 200-watt lights still throw quite a punch. I still have the scar where one hit me last week.

3. Black foil wrapped around a light will send its beam only where you want it. Use reflective foil (silver, white, or gold) to send that same shard somewhere else. Aluminum foil works in a pinch, but it's not as durable as the video kind (but more expensive). I bought 30 rolls of foil once to cover the entire ceiling of the location. It produced an even light, but it was a bear getting it back on those cardboard tubes after the shoot.

4. If you can't get a light exactly where you want it, try bouncing (the light, not you). Deep, dark crevices soon become happier places with soft, white light.

5. In order to get the shots you need, don't be afraid to take the camera where few have gone before. Your job is to get the best angle of what you're shooting and sometimes that means climbing the (corporate) ladder. Position yourself between the lights to get closer to the action. This may be uncomfortable because of the heat, but get your shot and leave (after you've finished getting your tan).

6. Foam core (next to gaffer tape) is your best friend on any shoot. The white board can be used as a bounce card as well as a flag. Can you say that about anything else on a shoot? You can say it, but would it be true?

What's so Swell about Gels?

First of all, I'm not talking about the gel you put in your hair. For the sake of the misinformed, I'll be talking about the little sheets of plastic that you put over your lights or your lens. That's right; you can actually put a gel over your lens and shoot through it, but we'll discuss that method of attack later.

In my opinion, gels are the most inexpensive way of changing and molding the quality of light on a shoot. With gels you change the look, intensity, and color of the light. To making your productions more professional in appearance, don't be afraid to experiment with that $8 piece of plastic.

How It's Done

Some people don't want to be bothering with gelling their lights; they just balance the video camera and shoot. But the auteurs might want a slight blue cast to a scene to signify coldness or a little orange to make the star look a little warmer and more glamorous. Some want that purple or pink cast to the shot; gels can make all of these people happy.

Probably everyone who's worth their salt in the video business has a swatch booklet of colored gels. They're fun to collect, amazing at parties, and really come in handy for, um, picking what kind of gel you want to use. Let's start there.

Use your swatch booklet for what it was designed, that is, as the best way to pick which gel is most suited for you in a given situation. Once you get past all the nifty colors, you'll find that each of them has a purpose, be it color correction, color enhancement, or color or light absorption. But the bottom line is that they all affect color and light. Diffusion materials or silks some people classify as gels, but I'll just be discussing color gels and not diffusion material.

One way to determine which gel works for you is to simply look through it. Basically, that's what the camera will see. Now try shining a light through the gel. Note what color the light gives off while shining it on a white card.

Don't take the manufacturer's word or anyone else's as to exactly what a specific gel will do. Try them in a practical situation. A tungsten light that's six hours old acts differently (and may have a different color temperature) than a new bulb. Just like people, gels also get old over time. Don't expect a gel that's been used for 20 years to perform exactly like it did when it was new.

A gel from one manufacturer (Lee) also can act slightly differently than one from another (Rosco). But what exactly do gels do? Simply, they give you control over light. A key source of light that's too blue or too orange can be corrected and made right with a gel.

Color Correction

Let's talk about color-correction gels. How do you know how much color correction a scene needs, if any? Most professionals use a three-color meter. This type of meter will read color as a red-blue shift and either a plus or minus green. If you don't have a meter, you can always use the other meter you have: your eyes. Don't be afraid to look through a gel; you won't go blind. The only way to tell if it will work is to look through it or use it on a light source. If the scene looks too blue, white balance or add orange light. The color balance of light that the meter reads is usually spelled out in degrees Kelvin. In school, we learned that perfect daylight is 5,600 degrees Kelvin and tungsten is 3,200. Always point the meter at the source of light you want to measure. Let's say you're taking a light reading in a room and it reads 4600 degrees Kelvin. That means you have a mixture of daylight and tungsten in the room. If you're not doing an avant-garde video, how do you correct for this?

One way is to grab your swatch book and hold a color correction gel over the face of the meter and take another reading. Keep doing this until you get the desired color temperature you're looking for. Once the balance is normal, add that gel to your light source.

Some cynics will say, "Why do I have to spend all that money for a meter when I can just white balance with my video camera?" You can do just that. A meter will only tell you what temperature you *do* have, so you know how far off you are and what you need to do to correct it. If you decide to use the white balance approach and still want a specific color to the look, hold the gel in front of the lens and white balance on a white card. If you do this, the entire scene will be balanced with this type of correction and you will have to put this gel on *every* light source.

If just the key or fill needs correction, don't balance with the gel over the lens. Remember that the gel should be the opposite of the color you want. For example, if you want a warmer or more orange look to your overall color balance, hold a blue gel up to the lens and white balance. Of course, this doesn't have to be full orange or full blue. Your special book has varying degrees of blue and orange. It even comes with a groovy name like *color temperature orange* (CTO) and *color temperature blue* (CTB). These gels also come in 1/4, 1/2, and full strengths. That way, if you want just a slight orange edge, you don't have to use full CTO; use 1/2 or even 1/4. By using this method, you will see on the monitor what the shot will look like once the gels have been added to the lights. These gels come in varying degrees (pardon the pun) of intensity. If a full CTO is too much correction, try a

1/4 or 1/2. To make life more difficult, gels come in booster colors. Booster blue does exactly what the name implies; it boosts or adds blue to the shot.

Once the shot is balanced, and you still want more color, try a bastard gel. Although these types of gels aren't really used for color correction, they do add a subtle color shade to the scene. So if you still want to add a splash of color, try a colored gel of that particular color (straw, amber, chocolate, and so on).

Speaking of chocolate, this is one of my favorite gels that has a specific purpose. When videotaping chocolate, use this color gel over the light. This really makes the confectionery more appetizing. I've also used it on dark wood backgrounds. Also, when videotaping dark-skinned people, the useful gel enhances the shot. Using chocolate gels makes the light seems to taste better; try it and you'll agree.

This all sounds like a lot of work, but unless you correct these light sources, you're going to have different colored pools of light in your shot. Experiment—that's what gels are all about. In video, you'll immediately see the effect a gel gives.

Just remember, you're in control of the light. You can make it look any way you desire. We have the technology. Can you over-gel (I'm still talking about lighting)? Yes and no. If you put three 1/4 CTB gels together, that will still add up to 3/4 CTB. This will work if you don't have a 3/4 CTB. But if you put red, green, blue, yellow, and magenta gels over the same light, you might have achieved the same effect with just one gel of a particular color (with a lot less light lost).

Being a Good Model

So what is a practical use for color gels? How about a lingerie shoot? For our shoot, the model was in front of a gray muslin backdrop illuminated with a 1,000-watt DP. This DP was gelled with a Rosco #44 middle rose gel that gave the backdrop a reddish, rose-colored glow.

The key light was a 2000-watt Molette. A two-by-four silk attached to a C-stand diffused the Molette. Two-1,000 watt Totas supplied the fill light. Each Tota was bounced off a piece of foam core. The only direct light on this shoot was the gelled DP pointed at the backdrop. The rest of the light was bounced or diffused to soften the look of the light.

When the model, Joy, arrived on the location we attached Rosco gold Cinefoil to all foam core surfaces. This gold foil warmed her skin tones and gave the film a more romantic feel.

Figure 4-5
Lingerie gel setup

Joy's outfit was sheer black and all the intricate design work disappeared against her skin. We had to find a way to increase the exposure on the lingerie without overexposing Joy's skin. This is where the flags come in. The best thing about Figure 4-5 is the great, attractive, colored lighting.

Right before the camera began to roll, the client wanted Joy's skin to glisten more. The gold Cinefoil had warmed up the color of her skin; now it had to look moist. We tried spraying water on her skin, but the drops were constantly running down her arms and legs. Because of our 95mm lens, the water looked more like sweat than glistening skin. Perspiration is a no-no in lingerie spots. The next step was baby oil. Baby oil makes the skin shine without forming droplets, another crisis solved. (I know they get olive oil from squeezing olives; where does baby oil come from?)

If it is your responsibility to light a shot, do what you want and need to do (using gels) to get the exact effect you want. A lot of this may sound like common sense, but experiment and find out what works best. Who knows, you may just discover that unique color! If you do, feel free to name it after yourself.

The key thing gels does for you is give you control of the light in a given situation, and who doesn't want to be in control? Once you start using color correction gels, you'll be hooked for life.

Review: Four Ways of Making the Beautiful More Attractive

1. Use colored gel on your lights to add bright hues to your set. The eye will be attracted to these rainbows (and so will magpies) and this is one instance where a multicolor set is okay. Besides, everyone will be looking at the pretty lights and not the lingerie (you can't believe everything I tell you).
2. Use gold foil as a reflector to warm the skin tones and silver to make them look colder. Depending on the look you're after, these foils add a nice touch to the shot. Unfortunately, this will probably be all you *can* touch on the set.
3. If you want the model's skin to glisten, don't use water because it looks too much like sweat (okay, perspiration) and will run off the body. Instead use baby oil, which is thicker and will stay where you put it.
4. Keep the set warm. If the model is happy and warm, that should come across on the video.

Lighting Ceilings

Challenges are what make life (not the cereal) exciting. One such challenge that I was recently given was to shoot a corporate video for a ceiling tile manufacturer. Of course, that wasn't the challenging part. The part that was most difficult for me was to illuminate and record a white patterned ceiling being installed by white, pale-faced people. The following is the saga.

How to Soften the Blow

The video proved to be very interesting and not complicated to shoot, except for than darn ceiling. It seems that the client makes over 20 different styles and patterns of tiles to be used in suspended or drop ceilings. Each of these tiles, planks, or panels can be installed using staples, furring strips, adhesives, spit, or a number of other ways. Of all the many patterns, lengths, styles, and types of ceiling tiles, you could get them in any color you liked, as long as it was white.

In fact, the patterns themselves were quite ornate. You could even install a white ceiling that looked exactly like punched tin (the type of ceilings that were very popular at the turn of the century). A type of ceiling could be made to suit any need, taste, or design.

My challenge was to illuminate these various types of ceiling tiles as they were installed, but the viewer still had to be able to see the different minute patterns and the grain that was etched in the surface. One of the patterns was just a 12-by-12 tile with thousands of tiny pinholes. And one more thing to remember, don't let the actor's hands blow out.

The only conceivable way we could light this type of setup was to use bounce lighting. My lighting director, George Winchell, came up with a different approach in creating the perfect bounce contraption. Using his MacGyver-type units, this type of lighting would enable us to soften the glow on the actor's hands yet still have the pattern emerge in the tiles themselves. Since this installation process would be shot in video, I had to shoot at a minimum exposure of F4. The last thing we wanted was to have the actor installing a large piece of blank white nothing on the ceiling. Figure 4-6 shows that white ceiling can be lit effectively and Figure 4-7 shows how it's done.

We finally came up with two slightly different bounce lighting apparatuses. The first unit, which was called the baby bounce, consisted of an Arri 600-watt Fresnel attached to the back extension of a C-stand arm. At the

Figure 4-6
Lighting ceilings

Figure 4-7
The contraption in
question

Figure 4-7
The contraption in
question

other end of the arm, a knuckle was attached to a three-by-four-foot piece of foam core. This arm containing the light at one end and a piece of foam core at the other was attached to a C-stand. This unit could be raised or lowered via the C-stand and pivoted using the arm. Like the hands on a clock, the baby bounce could be positioned from twelve to eleven o'clock, each "hour" giving us a different degree of lighting. Because of the weight of the light at the rear of the arm, the pivot point had to be tight or the hands of our clock would swing down to six o'clock at the most inopportune moment.

To keep the Arri's light fixed onto the form core, the unit was barn-doored slightly and spotted so all of its available light would be hitting the angled piece of white foam core. This unit was then situated about two feet from the talent and would act as their key light source.

Unit number two, the *big bounce*, had slightly more punch. Instead of using a 600-watt Arri, we used an open-faced 1,000-watt redhead. This unit was also barn-doored to allow most of the light to hit the foam core, but because of the lamp's wattage, we had to take extra precaution. Since more light spilled from the edges of the unit, another thinner piece of foam core was clipped onto the barn door. This clipped attachment would flag the light from spilling where it wasn't wanted and still act as a bounce card.

The actor would then climb a ladder until his head was a few inches from the ceiling. He would hold a piece of tile in his hands while talking to the camera. This piece was then installed.

The big bounce was positioned two feet away from the actor and raised to a height of six feet. This height allowed the unit to be moved closer to the ceiling if more illumination was needed, or lowered if his hands would start to glow. Because the lighting instrument was close to the ceiling, extra care had to be taken so the light still had enough room to vent without making our set a flammable one.

This big bounce setup worked well and was easily moveable. The soft, indirect bounce lighting made the actor's hands not literally or cinematically burn, it allowed the ceiling to be lit well and evenly, and most importantly it enabled the viewer to see if we actually were using a simulated white woodgrain tile or punched-tin tile, or even to see if our seams were straight.

Eight-tenths of this video took place three inches from the ceiling. The actor and I spent most of the time on ladders in an area where all the rising heat from the lights collects. At any given time, he could install a tile, point to its pattern, or run his hand along a seam without the exposure going nuclear.

The baby bounce accomplished the same purpose only with less punch. In fact, at times the big bounce was used as the key, and the baby, positioned farther way, as a fill.

This by far was the easiest way to shoot something of this nature. The lighting was kept consistent throughout the entire shoot and no one got sunburned.

Working on the Set or on Location

Some people say it's easier to light small objects or sets than it is to light a huge one. Does that mean lighting a larger set will be more difficult? Not really. Before I begin telling of the do's and don'ts of lighting a large area, I'll tell you what items we didn't have access to.

Of course, it would be easy to light a large area with Maxi Brutes, 10Ks, Musco Lights, or even a measly 5K. The largest unit with which I had to create illumination was a 2K. In this day, more size or wattage means more power!

Play Indoors: The Indoor Set

In the sequel to the corporate video I mentioned, the client had a large wooden two-by-four room created in the middle of a studio. Our job was to light this open studded set and show how the talent installed a ceiling. Since we were installing a ceiling, the roof above our wooden set was covered with plywood. This way the talent could install the ceiling over the exposed two-by-eight rafters. The bad news was we couldn't position any lights above the set. All our lights had to come in from the sides of the set. The two-by-four set described can be seen in Figure 4-8.

We began by using a 2,000-watt open-faced light as the key. This harsh lighting source had to be softened, or the imprints of the ceiling designs might have baked onto the talent. The light was positioned behind a sheet of diffusion frost (Rosco 216 to be exact).

On the other side of this light, a 2K soft light was placed. Because the throw from this light is diffused, no additional material was needed in front of the light. If both lights weren't needed as a key source (two 2Ks), one of the units could easily be positioned on the other side of the set if we happened to be facing that direction. These units, just being there on a

Figure 4-8
The special set

particular side of the set, enabled us to set up much more quickly. When you're in a hurry it's hard to unplug a light, wheel or carry it to the other side of the set, and make it operational.

In addition, having two 2,000-watt light sources next to each other doesn't make the light source a 4K. It makes it two 2Ks. One 4K actually puts out less illumination than two 2Ks. Don't yell at me; talk to the manufacturer.

The fill light on our set was a 1,000-watt light (usually a redhead or a 1K Fresnel). This lower-output light still had to be diffused, especially when the talent would be positioned close to the edges of the set. This stuff is also great as a sun block on the beach; it's less greasy than those lotions and oils, and it doesn't wash off in the water. Just take a set of it with you the next time you're out in the sun and cover yourself with it. Just leave a hole so you can breathe. The material of choice: 216.

One great asset about having a set constructed of two-by-four lumber is that we really didn't needed any extra C-stands, arms, or holders for the diffusion. At first, we thought that the only way to hold a sheet of diffusion material was in front of the light. But George, our lighting guru, came up with the "pushpin principle." By simply thumb-tacking the material to the outer surface of the wood, we had a built-in stand. It also proved helpful to tack up a full sheet of diffusion material instead of just clipping it to the barn doors. We would usually double up on the layers of 216. If you have a full sheet tacked between four consecutive pieces of lumber, the entire sheet becomes a light source. This broadens the illumination and gives you more control of the light instead of having it just attached to the light. If we needed less light, we moved the light farther away from the diffusion. If we needed more illumination, we could remove a layer of diffusion and move the light closer. Do you see a pattern in all this? (Not in the diffusion, but what I'm trying to tell you.)

Besides having a key and a fill light, all good sets deserve a back light (or kicker). The back lights on our talent were usually smaller units like 300- or 600-watt Arri Fresnels. These Fresnels could be spotted or flooded to give our talent the highlights they deserved. A little 1/2 or 1/4 CTO would give their hair highlights they would normally pay their hairdresser for. These lights were also placed at various areas around the set so they could be quickly called into action.

One of the greatest assets on a set has to be the lighting kit. What I mean by kit are the lights that come three or four to a case. You'll always need much more than one (like potato chips) and they come in handy when you have to move around a lot on a set. It takes far less time to move an existing light than to set one up from scratch.

So, by having sheets of diffusion material tacked up all around the set, we had soft lighting wherever we wanted. Our big, powerful 2Ks didn't seem quite as nasty and we enjoyed an even, thoroughly lit set. Of course, if every single lighting unit was exactly like I described, we would have what you might call a very flat set. We also created small highlights in the set with our smaller units. A precise streak of light here, a small shadow there, these are essential for any lighting job.

What have I learned? No job is too difficult if you just break it down into smaller pieces. Use the walls of your set (if you have them), and don't be afraid to soften what lights you may have in your arsenal. You'll never know how light the burden may be (sorry). At least I'm no longer afraid of letting my diffusion show when at the beach.

Reaching New Heights (Without Falling)

The most sincere form of flattery is imitation, and that's exactly what we tried to do. Since the *Titanic* movie was so popular, our client wanted to do a take-off for his video. His firm had just completed construction on a new, multistory building complete with verandahs, balconies, and walls of glass. From the second and third floor balconies of this building, the entire campus of the firm could be seen; it was quite a view. Many people aspire to climb three levels of stairs to look at mud, burlap, and construction debris.

The Outdoor Set: Location

The client thought that if he imitated the scene in *Titanic* where Kate and Leonardo are standing on the bow of the ship, people would stand up and take notice (and get a face full of salt air). That was our first problem, the client wanting us to create an effect from a $200 million movie with a slightly smaller budget. The 200 sounded familiar, but the million part was gone.

It's too bad none of the viewers of our video would be taking notice of how difficult that particular shot was to create in the movie. The big boys had a huge budget, digital actors and digital effects, a realistic time frame (that kept getting longer), and a large crew that knew how to create an effect like this.

We went through several different concepts to create the duplicate of this sophisticated effect, such as using a Steadicam, Jimmy Jib, or Helicopter mount before we came up with our eventual (inexpensive) solution: scaffolding. Without having access to any computerized digital effects, we had to cut back slightly on the concept. Instead of the camera starting in front of the talent, lifting, and flying back behind her, a simple zoom out was allowed because the "effect" of the wind in her hair on the balcony was all the client really wanted in the first place. It's really hard to understand why some people have to start out with grandiose ideas and never really intend to use them at all.

The area where the actress would be standing was much like the bow of the *Titanic*. Metal railings on the balcony came up to the talent's waist and could easily pass for railings on a ship. The "wind in the hair" scene could easily be recreated because we had at least 20 mph winds that day. If not, I'd blow really hard. Luckily, the client decided that the swoop over the talent's head would be a little too complicated and expensive. Again, he was holding back on us.

The effect or joke was to start tight on the talent's face. As the camera pulled back, it looked exactly like the scene from *Titanic*. The actress was wearing the same clothing as Kate, the camera was shooting slightly below to create the expanse of sky above, and we would be adding ocean sounds as well as a music library's version of Celine's song. We couldn't use hers because of a simple nine letter swear word: copyright. Hopefully, the viewers would recognize our joke as the camera finally pulled out to reveal a building instead of a ship's bow.

The chosen balcony extended from the building by four feet. The other option was shooting the talent on the roof, but just looking at it gave me a nosebleed. The scaffolding was erected from the ground up (as it should be). Four six-foot sections were attached to a wheeled base. The wheels rested on plywood squares, which served to level the shaky beast. Grass had not yet been planted on the site and the dry dirt was the most firm ground we could find. The client wanted the camera to be positioned slightly lower than the talent so she would be shot from a low angle and appear more dominant in the frame.

Plywood was also used as the platform for the camera and lights. Directly behind the camera, a two-by-six board was stretched across the length. This served as back support for me. When I asked the crew how they were going to attach wood to the side of the metal structure, they produced a roll of gaffer tape. It's always nice to know that your life hinges on a roll of $10 tape. With the rear board strapped securely, I could easily do a chin up and the board wouldn't come loose (I only weigh 40 pounds). Figure 4-9 shows the edifice where I spent my remaining days, and the lights in Figure 4-10 helped keep me warm.

Figure 4-9
Scaffolding

In fact, they strapped the wooden support so securely it took the better part of an hour for them to remove it. Once the scaffolding was erected, the lights and camera would be handed over from the balcony. This three-foot distance really taxes the strength in your forearm. I'm not really a wimp, but holding a 25-pound tripod out at arm's length would make me look like Popeye (I already like spinach). We could have strapped all the equipment to our backs and scaled the scaffolding, but a fly with an attitude might have knocked us off. I didn't want to be written up in the *National Enquirer* just yet.

Figure 4-10
The elaborate setup

The only way up to the top of the scaffolding was to climb it. If this was Hollywood, I could have been airlifted to the platform. Also, no one could ever sneak up on me while I was on the platform because any object touching the metal structure would cause it to shake. The higher the scaffolding is off the ground, the less stable it becomes. I had visions of having to spend a month atop this structure like some radio personality doing a marathon.

I was determined to get at least 50 takes of the action from atop my perch. With this elaborate setup, everything had to work perfectly. If I fell off the platform, I knew I would burn out in reentry. As a side note, it seemed funny that I always get the "high" jobs. I can't remember the last time I shot something on the ground. I guess this means I'm moving up in the world.

Since we would be shooting the entire scene outside, and the building was in the shade 12 hours a day, supplemental lighting was necessary. We used an Arri 1,200-watt Par HMI as the key light. This light was placed on the scaffolding with me, extended to a height of 10 feet and sandbagged securely. Any movement at all on the platform would create severe camera shake. I know this because I was having a fight with a horsefly most of the afternoon.

The HMI was silked with 216. When the light was illuminated and pointed at the talent, she appreciated me just setting the HMI on "warm" rather than "broil" without the diffusion.

Twenty-five feet above the ground, we stretched the AC cable for the HMI as well as the camera's umbilical cord to the deck. One of our hired chimps would constantly try to swing on these strung cables, thinking he was still in the circus, but fortunately he was the only one who could have set up the scaffolding that securely.

Behind the talent, another 1,200-watt PAR HMI was used as a backlight. An amber gel was used to slightly warm up the color temperature. This backlight was the amber glow Kate was bathed in as she stood on the bow. The tripod was also firmly sandbagged, as was I. This way if I ever fell off of the scaffolding, I'd be firmly planted in the ground.

An extreme wide-angle lens was critical. I needed to pull out far enough to reveal the effect, but the scaffolding was still only three feet from the building.

Sound was another problem because of the high wind speeds. The silk on the lights was constantly fluttering in the wind. Weighted down with eight clothes pins, the HMI was finally silenced.

After numerous takes where one thing or another wouldn't cooperate, we had achieved the desired shot. When I'm older I want a motion control camera, wind machines that I can start on cue, scaffolding that wasn't assembled by a guy named Thumbs, and enough space to plant my feet firmly to steady the shot. But we did the effect cheaply and that's why we were hired.

Review: Using Scaffolding Effectively

1. Make sure the structure is erected correctly. Every latch should be secured and every nut tightened (I'm not talking about getting drunk before you scale the apparatus either).
2. Ascend the structure without any equipment except a director's finder (if you lose a director, this helps you find one). Determine which lens setting you will use before hoisting the camera to the platform, rather than finding out you are three inches too close. Never attempt to move the structure when you're standing on it.

(continued)

(continued)

3. The lights should also be mounted to the scaffolding with you. This will place them close to the camera, helping key light the talent and keeping you warmer in the high winds.

4. Once the exact position has been determined, bring the camera to your platform. Strap or sandbag everything down. The wind is stronger higher off the ground and things can blow off, like my hairpiece and beard.

5. Once you are on the platform, make sure you have everything you need. Take care of necessities before you travel to these heights ("I told you to go before we left home") as well as carrying spares of what you may need. Batteries will die, tapes will end abruptly, something will soil your lens, and silks will blow away. Bring a boy scout with you and don't think about running water.

6. Don't move on the platform when shooting. Any movement telegraphs (a *Titanic* term) into the camera and it will wobble. Remain close to the camera, hold your breath, and shoot. Breakdance only on the ground.

7. Get as many takes as possible from your new vantage point. The opportunity may not prevent itself again, and this may be as high on the "corporate ladder" as you'll ever be.

8. Treat your crew with respect; your life may depend on it. If the grips have tails, it's because it helps them climb easier.

9. Be wary of brand-new, unscratched scaffolding. Its new look means it hasn't been used before and blood washes off easier than it does on a rusted surface.

10. Insist that the client look through the viewfinder while on the scaffolding. Now would be a good time to ask for that raise (before they fall off).

CHAPTER 5

How-to Videos

Becoming an Expert in Something You Know Nothing About

With how-to videos, your job is to teach someone else how to do something. In order to best make that a reality, you must have some knowledge about the subject. The final video presentation should appeal to the novices as well as the pros. It's simple to do a video on a topic you are fluent in. Why do I always get asked to do videos on something I know nothing about?

Full of Hot Air

A client decided he wanted an aerial view of his manufacturing plant. He wanted the footage to be wide enough, clear enough, and steady enough (some clients are just too demanding) to use in a recruiting video. We discussed the options of using a jet plane (a 133.2-millisecond shot), a prop plane (not very steady and quite noisy), a helicopter (very steady, very noisy, and very expensive), and a hot air balloon (very steady, not noisy, and reasonably cost effective). It seems our decision was easily made.

When shooting in a hot air balloon, find the smallest, lightest camera you can find. If we were about to crash, I could always throw it overboard (that term works in ships and balloons). A larger camera would add extra weight to the balloon's basket. If I had fallen out of the basket with a heavy three-chip camera, I might have burned out in reentry.

Since all hot air balloons belong to an elite group of balloons, they have to leave at dawn—o' dark-thirty to be exact. We arrived at 6:45 A.M. and the balloon was already being inflated. Nick, the owner of Balloon Port, supplied the red, white, and blue neoprene vehicle for our pre-dawn flight. Twelve hundred feet of new fabric had been sewn into the balloon at a cost of $14,000. The deflated balloon weighed in at a whopping 230 pounds.

A large, high-velocity fan inflated the unfolded balloon in a few minutes. Since you can't really steer balloons, I was going to go straight up to a height of about 400 to 500 feet, get my shot of the complex from the stratosphere, and then come down before I got struck by a meteor. Nick informed me that once aloft, the air currents would determine our direction and speed. Since our flight was on a late October morning, I donned my sky jacket, grabbed my camera, and climbed in the basket. Figure 5-1 proves that man was not meant to fly.

Climbing into the basket isn't as easy as it looks. You have to stick your foot into a small hole the size of a quarter that helps you get into the basket.

Figure 5-1
Hot air balloon

Since my five toes aren't pointed or webbed, I tried unsuccessfully to perch on the edge of the basket (à la Snoopy) and propel myself into the unit. If I had a larger camera on my shoulder, I would have snapped in half. How do 80 year olds get into a balloon? Come to think of it, I never actually saw Nick get into the basket; maybe he just lives in there. Once inside, I tied my shoelaces together so I wouldn't trip over anything.

After a yank on a tether, the seasoned pilot Captain Nick had us ascending rapidly. The liftoff was incredible: smooth and silent. The roar of the propane-fueled burner was the only sound I could hear. Immediately, my glasses fogged up and I looked like Little Orphan Annie on a bender. I quickly stashed them in my jacket and continued to tape my assent. I also had enough battery power in my jacket to illuminate Toledo through Christmas.

It was almost as if I were using a Steadicam. The assent looked as smooth as glass (cliché corner). I was manually adjusting the iris because the sun was beginning to rise in the East (exactly where it's supposed to). So I guess I have to admit the smoothness of our rise was because of Nick and not my camera work.

As a side note, a balloon is the smoothest and cheapest way to get an aerial shot of something. Every other device has more bumps and expenses involved.

Because of the early hour and cold temperature, a thick covering of fog enveloped the trees and clung close to the ground like warts on a frog. The buildings were quite visible below, but everything living was creating a carbon dioxide haze. In the viewfinder, a one-chip camera has a harder time cutting through the haze. The detail just isn't there. Those two extra chips and hundreds of thousands of pixels provide the separation that's really needed. But, hey, the one chip was a lot lighter.

In order to insure my satisfaction, Nick kept asking me if I was high enough to record the entire plant from the sky. I said that I still needed to get higher in order to view the whole complex. I didn't realize at the time how fast we really were climbing. With one eye, I was watching our captain to see if he was pulling out the oxygen tanks. A bird flew by us on our way up, looked at me, shook its head, and dropped out of sight. I told Nick that if I could look into the window of a passing plane, we had gone too high.

At 1,400 feet, I was getting the exact angle I wanted. The air seemed awfully thin and my ears were popping. Because I was constantly staring into my little black and white world, I hadn't noticed how cold it was getting at that altitude. I could feel the water vapor from my breath icing up on my beard. I felt like I belonged on the cover of *National Geographic*. From our height, every dog in the planet was barking at us. I didn't think that the sound would be that clear at that height. Nick told me that dogs always bark at balloons because there's something in the propane burner only they can hear. I kind of thought that they were barking at my icing beard, but I guess I was just thinking of myself again. Nick also said that cows and horses are affected visually by balloons, and dogs and some small children by the sound.

After I had shot everything I needed and was starting to look for something to eat, I was ready to go down. Carefully watching for power lines, we descended. I could actually hear shutters on cameras clicking and squeals of children as we swooped down over rooftops. I had my 15 minutes of fame. The landing was a little rougher than the liftoff. By bending my knees when we landed, I didn't end up in some Chinese family's kitchen table on the other side of the globe.

Once the basket hits, it wants to drag along the ground because the balloon is still inflated. Being a novice, I tried to climb out the split second we touched down. With one leg halfway out of the basket, I expected to be dragging my foot for the next three miles. I would be a candidate for a peg leg. Nick pulled me back in and we soon stopped moving (I hadn't paid him yet).

So, if you need to get a hot air balloon shot, it truly is in the hands of your pilot (give Nick a call). Go with the lightest camera you can comfortably hold, although the new three-chip digital cameras would easily fill the bill.

Review: Five Tips for Shooting in the Air

1. Depending on what kind of aerial shots you want, decide which flying device works best for you. If you need slow, extremely smooth, steady shots, a hot air balloon is the only way to travel. If you need faster movement, banks, or reverse angles, a helicopter is better for these types of shots (but more expensive). If you want to shoot at six Gs, then an F-16 is the choice you should make (call me and I'll shoot it for you).

2. In a hot air balloon, keep your movement in the basket to a minimum. If you reach into your pocket for a tissue, the basket and balloon will sway sharply. The basket is normally steady and smooth and any bumps you feel are caused by you (unless you happen to hit a duck in flight).

3. The higher the altitude, the colder the temperature no matter what time of year you fly. Dress accordingly and make sure your camera is also prepared because the battery charge will drop quickly. If you start to lose altitude, don't offer to "help blow up the balloon." The pilot doesn't know you mean "inflate."

4. The haze on the ground is heaviest in early morning flights and video will not cut through this; it just looks out of focus. Try to plan your trip towards midday when the haze is gone and your balloon's shadow is directly beneath you. If not, you'll have to frame it out of the shot. Don't zoom from the air; it will make you and your viewer nauseous. Anything you toss out of the basket will burn out in reentry.

5. Although the flight will be smooth, when you return to the planet's surface you're in for quite a bump. Bend your knees as you land and make sure you have a copy of your will in your pocket.

Lower than Your Feet

How often can you say you've shot most things? After shooting roofs, ceilings, and walls, all that was left to shoot was flooring. A client hired us to videotape a how-to flooring installation. A new model home was chosen and our job would be to transform the *orchestrated strand board* (OSB) subfloor into a rich gunstock oak laminate floor.

I can honestly say I learned a lot on this shoot. I had never worked with laminate flooring before so that was where my journey began. Laminate flooring is made from compressed wood pulp and saw dust. This material makes up the quarter-inch-thick backing (and the strength of the floor). The wood grain on the surface is actually a photograph of wood that is laminated (thus the name) under a protective plastic coating. This photographic wood is the thickness of a piece of paper. The part of the laminate flooring you see is actually the thinnest part.

Using this process, the floor can look wonderful. The grain patterns always match and each piece looks perfect. The most amazing thing about laminate is its durability and ease of assembly. But for now, let's tackle the challenge of lighting the scene.

The OSB subfloor is the usual subflooring in a new house. Laminate flooring can be installed over vinyl, concrete, or plywood. Our model home had drywall-dust-impregnated OSB. The walls and ceilings (drywall) were painted white, so actually the room was one big bounce card. This would help immensely in our lighting plan. Figure 5-2 shows how to best shoot a flooring video.

Our key light was a Chimera with two 1,000-watt Totas. The white fabric in the front of the Chimera gave us soft, warm, diffused tungsten light. The fill lights were 1,000-watt DPs. The open face of the DP was bounced against either the ceiling (fill light number one) or against a piece of foam core mounted on a C-stand and arm (fill light number two). Therefore, all the lighting in the room was bounced.

Figure 5-2
How-to flooring

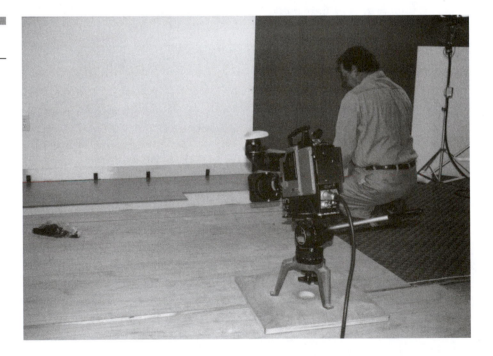

The window in the dining room (our first installation room) was gelled with 85ND9 so we could leave our tungsten sources ungelled. I usually like to work with a minimum F4, but the lighting and white surfaces gave me an F5.6.

As the installer began laying out her flooring material, we determined that the dark gunstock oak flooring would need more illumination or it would appear too dark on camera. Our grip, Brad Kenyon, produced China balls. China balls or Chinese lanterns are manufactured by Chimera, Lowel, and a number of other sources, but our $4 paper specials were from Ikea. The only real difference between our lights and the professionals were that ours had open bottoms and weren't completely enclosed. This created one drawback.

In the cutting room where all the laminate flooring was cut to size, we used the Chinese lantern as our only fill light. The floor-to-ceiling picture window offered gallons of daylight so we opted not to gel the window and balance the camera for daylight. The installer would have plenty of natural light without the heat created by tungsten lights.

In order to soften the harsh shadows created by the ungelled window, we used a Chinese lantern with a daylight-balanced (blue) photoflood bulb. This 300-watt unit was hung from an extended C-stand arm over the installer's cutting area. The 360 degrees of soft light filled in all of the shadows and gave us a very pleasant look.

However, since the lantern was an inexpensive unit, the photoflood's light shown through the bottom of the China ball and cast a circular ring on the work surface. This unnatural source was extremely distracting and was soon covered with gaffer tape. The light was still allowed to vent out of the top of the unit. If not, we would soon have a very flammable light source. Although not a disease, China balls can be seen in operation in Figure 5-3.

In every shot for the rest of the installation, the Chinese lantern was used in the dining area; the 300-watt daylight-balanced photoflood was swapped for two 150-watt tungsten-balanced photofloods. The two lanterns provided the same type of fill as the single unit.

When working on the floor, the lighting must illuminate the action. The foam core was tilted down and the DPs were pointed up at the foam core. This bounce light provided a wash on the floor and brightened the skin tones of the installer. The Chimera was pointed down to provide additional key. Remember to keep the rear flaps of the Chimera open. The heat from the enclosed Totas can fry your AC cables rapidly.

The laminate floor extended from the dining room into the hall and a closet that housed the water heater. In order to illuminate this tiny closet, we had to install a light in the crawlspace and work upwards.

Figure 5-3
China balls

For some reason, green drywall (which is water resistant) was chosen for this closet. Every other drywall surface in the home was white, but this space was green. Somehow a light bouncing off this surface gave the floor an unhealthy green cast. Foam core was taped to the angled surface of the wall and our light was once again white. In this cramped space, a 750-watt Omni was used as the light source. Extended on a C-stand arm, the light was placed one and a half feet from the foam core.

Once the flooring had been installed, we had to show how easy the cleanup was. Glue from an assembly that is left on the floor can be removed by simply pulling it off once it dries. This is easy to say but difficult to show. The glue applies white but dries clear. How do you show a transparent piece of glue on a dark wood floor? The only way we could figure out was to skim the surface of the wood with light. A 200-watt Pepper was attached to a C-stand arm and lowered to the floor. The 200 watts were too strong and the light had to be diffused. With a piece of 216 attached to the front, the Pepper became our glue-finding light.

Held at a 30-degree angle above the floor, the clear-dried glue could be easily seen by the camera. By scraping the edge of the glue, the end could be lifted and dried glue removed. We were so excited by this technique that we created 10 test patches of glue and watched them dry (okay, it was a slow day).

The key to lighting flooring is to create a soft, bounce light with enough foot-candles to bring out the texture and grain in the wood without overexposing the talent. A foot candle is the mount of light one candle gives off at a distance of one foot. 60 foot candles would be the amount 60 candles give at 1 foot. The white walls of the room also helped in the softening effect. The next time you're asked to shoot the floor, remember a soft touch is better and leaves no footprints.

Review: Five Tips for Shooting on the Ground

1. If shooting dark flooring, increase the light against the surface. Darker wood absorbs light and lighter surfaces will reflect (that's a good reflection on you). Don't point the light directly at the floor unless you want hot spots, but increase the intensity to offset its darkness (that was my previous boss's title, "its darkness").

2. Use soft, nondirectional lighting like China balls to evenly fill out your location lighting. The inexpensive Ikea models work just as well as the professional units and at a few dollars a piece can be thrown away if damaged (I can say the same thing about our key grip).

3. Try not to mix daylight and tungsten sources unless you want colored pools of light. Determine the strongest light source in the room and gel accordingly. The same principle applies when combing your hair after a shower.

4. Place the light at an angle if you want to highlight objects or imperfections in the floor (or look at grain patterns). The flatter the light, the less detail will be seen. Why do you think female models are always lit at an angle? They call it modeling; I call it professional.

5. Use soft, bounce, or reflected light on the floor. This eliminates shines, shadows, and unevenness. This same thing applies when shooting bald people.

Reaching the Entire Audience

Since you have the power to make videos, you are controlling what the viewer or audience sees. The goal is to make everyone watching your video love it beyond belief. Of course, this means that something in your program must suit everyone. If you talk above, below, or against someone about something, you're going to lose them. How do you please everyone? You can't. How do you please most of the people, also a tough call, and make yourself happy (the easy part)? There must be something in your video for everyone.

I thought shooting food would have something to make everybody happy. Everyone's got to eat, so show different types of food and how they are made to look pleasing for the video camera. Once again, no bubbles will be burst, but you can learn how to make low-budget food look appetizing to reach the entire audience.

Don't Be Afraid of Make-up

I know a lot of people who wear make-up and a few animals, and now I've learned that food can also wear make-up. I'm not really talking about the kind of make-up you put on with a powder puff, but other elements that make the food look good under hot lights.

A client happened by who wanted a cooking video produced. He told me that because he was a gourmet, his food was the tastiest on the planet, and it had to look that good on TV. I told him that unless the food showers and shaves before being on camera, it wouldn't look very good. When my joke flew over the top of his head, I explained that I was talking about food styling. I told him I knew a top-notch stylist that would make his food be better dressed than the people eating it. He agreed to hire the stylist and the following are some tricks you can use to make food look better when in front of the camera (and save the stylist's salary). A table full of food with make-up applied is shown in Figure 5-4.

Hours before the shoot began the stylist set up his table of goodies. Most of the things on the table you'd never associate with food, but this fellow used everything on the table.

We incorporated stand-in meat until the lighting had been fine tuned. Stand-in meat can be consumed; real meat in videos can't be eaten until it says so in the contract.

Figure 5-4
Food wearing
make-up

The first shot was going to involve crab meat. After the shell has been removed, the white lump of meat doesn't look that appealing. It must be made up. After lightly browning the meat, the stylist used tweezers and later an X-acto knife to remove fine hairs or "fuzzes" on the meat. I told the client earlier that the food had to be shaved. Various ridges and valleys in the meat were pulled and poked to enhance the texture of the product. In order to add a slight brown color to indicate crispness, gravy coloring was brushed onto the top of the meat (this is the dye that comes in a bottle that people create gravy from). The top of the meat was brushed with olive oil to make it glisten slightly.

Over the top of the table, a six-by-six silk was erected to soften the key light. An Arri 2K provided more than enough illumination without overly baking the food. To camera right, a 250-watt Pepper would provide the fill light, supplemented with a 600-watt Omni. Mirrors were also placed just out of frame to reflect light back onto the surfaces of the food. At times, at least six of these make-up mirrors supplied just the right amount of highlights in the crevices of the food.

Very little food styling has to be done to crab meat when you are a half-inch from it. In fact, at that distance it's really hard to tell exactly what you're looking at. It could pass for cauliflower, my Aunt Ethel at the beach, or chipping paint. In order to give the viewing audience a clue of what they really were seeing on their 32-inch Trinitrons, we placed the crab on a turntable (I removed the Barry Manilow record first).

With a rheostat to control the speed of rotation, the camera remained stationary as the crab breakdanced for us. At a slow crawl (a crab term) rotation, you were actually able to tell what you were seeing and still get close enough to go where no one had gone before.

Without the use of motion control, it was nearly impossible to pull or push the dolly slow enough to get a movement without jerks, stops, or starts. No matter how many times we tried, we just couldn't get the movement smooth enough. As anyone knows, hovering is smooth, not bumpy.

The client suggested we wedge a Matthews C-Stand arm between the dolly's pulling bar and the dolly track itself. With a slight push or pull, we now had a fulcrum in which to pivot the dolly. With a slow lifting action, we could push the dolly a total distance of four inches, just enough for the length of the shot.

As the camera recorded the movement, we started with a full screen of meat and ended 10 seconds later still showing a full shot of meat. It was as if a very nearsighted person were carefully examining his meal choice for any defects.

The potato sitting next to the crab had a similar beauty treatment by being brushed with olive oil and spritzing the skin with soapy water (to create glistening water droplets). Right before the cameras rolled, the ice-cold, raw potato had to look hot (cooked). This is where AV smoke came into play (purchased from any AV supplier). By dropping the A solution (SE-502) into the open potato with a straw (capillary action at its best), the potato would smoke when the B solution (SE-503) was added. If we wanted more steam, more B was added to the A.

If we lost sight of the food through a haze of AV smoke, we'd add slightly less B to the chemical mix. Mr. Science would have been proud of how we chemically created an inviting, appetizing, stone-cold potato with nature's chemicals. The food looked good enough to eat, but we'd probably die if we ingested the chemicals on the potato.

It would be rare if anything would be left alone after the stylist finished with it. There were always final picks, pulls, probes, and finesses before the tape rolled. Everyone had to be happy with the look of the food, not just the stylist. A little piece of fuzz looked like a giant log. Talk about getting intimate with your food!

Stop Whining

The white wine behind the plate of food also had to be dressed. It was real white wine from a freshly opened bottle, but the client just didn't like the way it looked. I couldn't figure out what he could possibly do to it to make it look any more inviting. That's when the stylist dropped in a plastic ice cube. It still looked good on the rocks, and the fake ice would never melt. The client still wanted more. With a wooden skewer dipped into liquid soap (dishwashing liquid), tiny, frothy bubbles clung to the edge of the wine glass. This soap created a film on the surface of the wine that had to be constantly stirred (James Bond wouldn't have liked it). But I must admit the wine looked better. Just don't drink it!

The whole Cornish hen experience I'd rather not discuss, but it was a trick I never would have dreamed of doing. The hens were cooked slightly until they were brown, but they had to be really brown on camera. Not having time to send them to the beach for a week, the stylist used Miniwax stain and stained the hen. The walnut stain remained on the hen's skin and gave it an appealing if not flammable hue. Once the stain dried, more olive oil was added for the sheen (I was surprised that shellac or varnish weren't used to complete the process). Figure 5-5 shows naked food (photo places

Figure 5-5
Food getting dressed

Figure 5-6
Crab really close-up
and mirrors

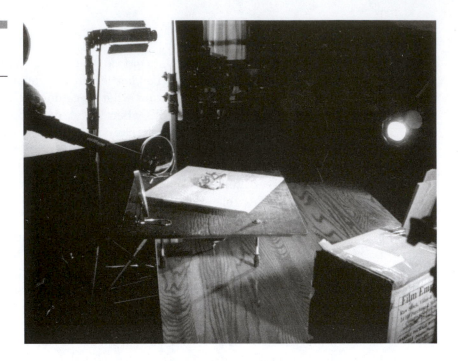

will still develop pictures like these). If you ever really wanted to get close to a crab, you may have your chance in Figure 5-6.

While one product was being shot, the next item was in the green room awaiting make-up and learning its lines. The snow crab legs didn't need any make-up because they looked entirely different cooked than raw. The client wanted them to be lifted out of a steaming pot to signify that these legs were freshly steamed. We tried our AV smoke, but it wasn't enough and the smoke had a hard time rising to the top of the caldron. The stylist used one of the oldest tricks in the book to create this steam. Place four or five big chunks of dry ice on the bottom of the pot and add warm water. Since the dry ice is so cold (something like 160 degrees below zero), the warm 80-degree water creates instant smoke. In a shallow pan, this would never work because the dry ice's fog has a tendency to roll over and off the edges of a pan. It looks too Addams Familyish. But the high sides of the caldron let the fog rise slowly and mist into the air. At least we could eat these legs after the shoot.

The last seafood shot was lobsters. Like crab legs, they look red when cooked so they had to be prepared. However, when they were garnished with butter, they looked like a freshly waxed Porsche. The unwaxed lobster wasn't even worth looking at compared to the on-camera model.

The rest of the food being shot employed the same techniques I mentioned earlier. Hair dryers, hand steamers, various surgical appliances, liquids, dyes, Q-Tips, cotton balls, and plastics were used to dress this food. I can't think of a better way to stay on my diet. Now I'm afraid to eat anything that hasn't been to the beauty salon.

If the minimal budget allows, rent a snorkel lens to get extremely close to the food. The client will see shots like he or she has never seen before.

Review: Ten Things to Do When Snorkeling

1. When renting a snorkel lens, get the full array of prime lenses to use (all five). When I thought a normal 25mm lens would work, I was way off when it was attached to a snorkel. Maybe that's why fish think I look weird underwater.

2. Experiment with and without the angled adapter. Sometimes sliding along just below the food is better than flying in from about. Because it has so many screw on attachments (just like me), make sure each diopter or lens is seated properly before you screw it in. No way am I going to make a joke here with that previous statement.

3. When using a snorkel lens, you will be getting closer to food than you ever have before (unless you're Mr. Magoo). Make sure the food is dressed properly before shooting it. Just try to say that to the head waiter the next time you dine out.

4. When using the diopters, it's just like shooting in macro (extreme close-up); every move is magnified. Keep the camera tripod mounted and never attempt to handhold this rig. You will poke someone's eye out.

5. If the shot involves movement, use the smoothest dolly you can find. Moving just two inches with a snorkel seems like a few feet on camera. Gaffer-tape your hands to the dolly because the sweat rolling off your palms will also cause the camera to move.

6. This snorkel lens is *long*. Just watch where you're pointing that thing because many times I rammed into the food because of my extreme length (of the lens). The food stylist didn't like filling in my lens poke holes.

7. Make sure you balance your tripod correctly because the snorkel adds a few pounds up front. My wife has said the same thing about my stomach.

(continued)

(continued)

8. When you finish the shoot, have everyone's picture taken with the snorkel lens. The lights, crew, and camera attracted no attention on the shoot; everyone was fixated on the lens. Now I know how Pinocchio felt (and I wasn't lying).
9. Try to shoot with the highest F-stop possible. If not, your depth of field will be almost nonexistent. Of course, the food won't taste as good when shooting at F64, but at least you'll get a tan in the process.
10. Rehearse everything a thousand times before you shoot. This lens (okay, because it's so long) is like nothing you've ever used before. Treat it with respect and don't try to use it as a light saber with the key grip after the shoot.

Fierce Competition: Is It Really?

Since everyone and their uncles are making how-to videos, how can you compete with them? The easy way is to make your video better than the others, and since you have a lower budget then they do, you may have to work a little harder.

Shoot Golf In or Out of Season

Unlike most sports, golf has no equal when it comes to videotaping. I know it takes a quick eye and a steady hand to shoot any sporting event, but golf is more challenging; golf balls are smaller.

I have videotaped my share of skiing, football, and track events, but I really wasn't prepared for what I experienced in golf. I figured it would be an easy shoot because not much happens with the ball. Golf balls are just hit with a stick, or they are kicked, thrown, or bounced. Golf isn't a contact sport like football when the players are constantly moving and the ball is difficult to follow. I didn't think anyone would even break a sweat while playing golf. I really had a lot to learn.

My only experience in playing golf is trying to get my green, red, yellow, or orange ball in the clown's mouth for a free game. The real golf the pros play is a lot different. I had to be led around by my nose.

A private country club in Maryland wanted to shoot a golf pro for a how-to video. There aren't any videos on how-to golf being produced in our area, so I figured we had the market cornered with no competition. We would spend the day shooting the pro and various amateurs playing the sport (I don't qualify under any category).

Because golf is an early morning sport, we had to be at the site and ready to go by 8 A.M. The sky looked threatening and we were prepared for anything. I had bright yellow plastic rain gear, a large plastic baggy for the DV camera, three batteries, and two cases of tape. With a 40-minute tape load and 18 hours of batteries (if I didn't opening the side-viewing screen), I could tape everything you ever wanted to know about golf.

In the very beginning, I was way off base when I noticed how difficult it was to keep up with the pro. The pro and all the players at hole one were all over six feet four. With my shorter five-foot-eight frame, every player looked massive from my lower-angle perspective. With a quick swing, the ball disappeared and there was no way I could follow that tiny white dot in the viewfinder; it would be the size of a pixel.

I taped each player from several different angles. After each player had his turn at the helm, we got in our cart and flew to the middle of the green (actually everything on the course was green, so I was confused from the start). As I was taping a pour soul trying to get his ball out of the sand trap, I was nearly beaned by an errant ball that someone else was knocking around. I then learned that players only shout "Fore" to warn other players, not camera people.

Outside at a sporting event like golf, it's best not to use the auto features of the camera. The auto focus can't keep up with the tiny ball, and the auto exposure always has the players in silhouette.

As the day progressed, I got behind schedule. I really think the players were taking too long trying to get the ball into the hole, while we were working our tails off.

We always had to get to the players' balls before they did, so we could see their approach and their swing. The hard part was finding this white dot in a sea of green before they got there. Twice I had to silence caddies that found the ball before I did. I would usually distract them by showing them the camera as I stepped on the golf ball with my foot to bury it and halt their search. I also learned that it's important to frame the player's swing and not worry about following the ball after they propelled it into the stratosphere.

Every time I zoomed into a close-up of the ball rolling along the grass into the hole, it didn't. Whenever I chose a wide shot, that's when they would sink the ball. Every single shot was backlit on the golf course. The manual exposure would have to be opened to allow detail to be seen on the players and the background would blow out. The players all had this steamy, sultry look about them as we roamed the fields in 90-plus-degree heat. In the distance I could actually see the heat rising.

As the day worn on, I began to wear out. My glasses were constantly steaming up from the humidity as I tried to peer into the 1/2 square viewfinder. My flesh was searing because I had no protective sunscreen, and all the moisture from my body was long gone. As our cart skidded to each stop, I would dismount, run to set up my shot, and hope the golfer would play into my frame. None of them ever did what I expected them to do (especially the pro).

Luckily, every other green had water available for the players. Not only did we have to beat each player to their balls, we also had to get to these oases before they did. If we arrived late, they would fill their bottomless "big gulp" cups and all that remained for us was to lick the condensation from our glasses.

I had been instructed to get high fives, hoots and hollers, and shouts of "all right," but all I videotaped were sweaty people having bad scores. Being a novice to the game, I had to be told what each person had to do. I knew the little ball had to fall in the hole, but who went first each time, how they marked their balls, whether a player was a righty or lefty by looking at their clubs, and when to determine if I had enough footage of a particular player, were all beyond my comprehension.

As the day wore on, our cart ceased to function. As we traveled up a steep incline, the cart shuttered to a stop. I got out and pushed as my driver planted his foot on the gas pedal and pushed with his other foot. Slowly, the cart went up the hill. When the roller coaster started to go down the next hill, I had to dive into the cart or be dragged for the next hundred yards.

We were fortunate to find an abandoned cart and use it to push our cart up the next hill. Because I was lighter, I sat in the dying cart as my assistant would come up from behind and ram into my cart. Like being in bumper cars, I was whiplashed all the way back to the club house. That's when the skies opened up with a deluge of rain.

I did learn several important lessons on this outing. Shoot everything as wide as possible; this is the only safe way to capture the action when you have no idea what's going to happen. Use the camera's manual features; the auto controls have no better idea of what's going on than you. Don't use the flip-out screen because it's impossible to see in the bright sunlight and it sucks the life from your battery. Finally, prepare yourself for exposure to the elements, drink fluids, wear sunscreen, and, most importantly, learn a little about the sport you'll be videotaping.

Review: Five Golf Shooting Tips the Pros Don't Know

1. Don't use the automatic features on your camera no matter how sophisticated they may appear to be. The little ball you're trying to find is far too small and you have no control when the camera "hunts" to find its correct exposure or focus. Besides, this gives you something to do when waiting for the ball to fall out of the sky.

2. Don't try to search for the ball while looking through the viewfinder. It's the size of a pixel and if watching it in black and white, you might end up following a small cloud rather than the ball. Instead use your good eye. Too bad we don't have crosshairs and zooming eyeballs like Steve Austin, the *Six Million Dollar Man*.

3. Get as many reaction shots or cutaways as possible in any sporting event. You will lose sight of the ball (and life sometimes) and players staring into the stratosphere might help. Don't use silly music when they can't find their balls.

4. Shoot everything as wide as possible. When you locate the ball, you can always zoom in to a close-up. A zoom in is more appealing than a snap zoom out and a flash search for the item. This way you always have the ball in the shot if the viewer can see it or not.

5. If equipped with a flip-out screen, keep it closed because the glare and battery drainage will make you unhappy. If you need to check a shot quickly, the flip-out screen can be used in a pinch, but you will see more detail outdoors with the viewfinder monitor. I've also been known to flip out on shoots without even using the screen.

Two for the Price of One

There usually comes a time in your life when something you're shooting is difficult or expensive to repeat. The best thing to do in situations like this is to shoot the action with two cameras. When Hollywood sets up a difficult stunt in a film, you can be sure the action will be captured by more than just one camera.

On another "how to install flooring" shoot, two cameras were certainly better than one. When showing the viewer step-by-step action, the camera setup has to change so the viewer can see the same action in a close-up. Because our flooring was glued together, our installer would have yelled if he would have had to pull apart his freshly glued floor for us to get a close-up of the same action.

Whenever you're watching someone do something in the hopes of learning the skill yourself, you want to see the action from as many different angles as possible. If you're there in person, you'll walk around them and examine the activity from different angles and as closely as you can.

With two cameras called into action, one camera could focus on the wide or master shot, and the other camera could concentrate on just getting the close-ups. Of course, this doubles the cost because you have two cameras and two people, but it may cut your time by more than half.

The first shots were installing the linoleum flooring in the hallway. As the installer placed the fabric down, he then proceeded to glue the edges for a permanent bond. Camera one captured the master shot as he finished gluing the edges. Camera two recorded the close-ups of the glue flowing out of the bottle, the bead along the edge of the floor, and the glue oozing out at the edge. Because both cameras were shooting such widely different shots, we didn't have to worry that both cameras might be capturing the same action.

Much like recording a program on live television, in editing we could cut to the close-up at any time because the action was exactly the same as recorded on both cameras. If you ever watched *Cheers,* you will notice that every shot matches perfectly from the close-up to the wide master shot. The producers shot this program with three cameras so the action could always be matched.

If we tried to rerecord the event in second take (shooting only close-ups after we finished the master shot), the action would have to match perfectly or the continuity would be lost. This is very difficult for nonvideo savvy talent. Also the areas for cutting are much more limited with this approach. Obviously, a lot more tape is used with this two-camera configuration, but you'll spend less time installing and less time in editing.

It also helps to have both cameras have the same time code. Once again, in editing this identical time code makes it a snap to find the close-up when viewing the master shot's footage. Sometimes it's better having technology work for you. Figure 5-7 shows the placement of cameras for a two-camera shoot.

For all the action in the hallway, the cameras were positioned side by side at the same height level. The close-up camera had a lot more work to do following the action more closely, but we were able to complete the hall installation in less time and have much more coverage. This technique allows the director to cut the finished program any way he or she sees fit. Five different people will edit the same program five different ways with this

Figure 5-7
Laying big flooring

approach. The editing game plan should be discussed well before the cameras begin.

In order to light the minuscule hallway, we chose to bounce a 1,000-watt Lowel DP off the white ceiling. This bounce light filled the room with an even, soft light. Our next trick was where to put the light in the hall without seeing the stand. When you're showing a floor being installed, the light stands have to hover or they'll be seen.

When we moved on to the bathroom, the installer laid flooring from the doorway backwards, working toward the bathtub. Since both cameras would actually be in the doorway shooting down, we wouldn't see the bathtub. That porcelain unit became the home for the DP.

With the light carefully placed in the tub, another DP was used behind the cameras to illuminate what the installer was doing. When the tub light was turned on, the room was bathed in a warm glow. I'm just glad we weren't using a black bathtub.

As the installation came closer to the bathtub, we needed to find a different spot for our DP. With the 1,000-watt unit attached to a clamp, we fastened it to the shower head. As long as no one took a shower, we could use this obstruction to point our DP skyward without seeing any stand legs. Once again, a white tub works best.

At all times both cameras were on the same side of the talent. If we ever decided to switch sides, we would have to make sure we didn't cross the axis. I've heard about this place (the axis, and not crossing it) all through my career and have yet to actually see it. Maybe when I'm older.

During the course of the shoot, you can get very chummy with the talent, sharing such a confined space as the bathroom. At times, both of the operators had to be in the bathtub to get that special angle we were looking for. That's when you really know you love this business. You and a comrade are both sitting with your cameras and tripods in a bathtub, fully clothed, and you have a 1,000-watt open-faced lighting instrument above your head. It's best not to attempt this on a Saturday night; your bathroom is much too crowded then. As a child, I never wanted to take a bath; now I spend an entire day in one with a camera and a friend. You can share the tub with a friend as long as both of you are fully clothed, as shown in Figure 5-8.

Figure 5-8
In the tub with a friend

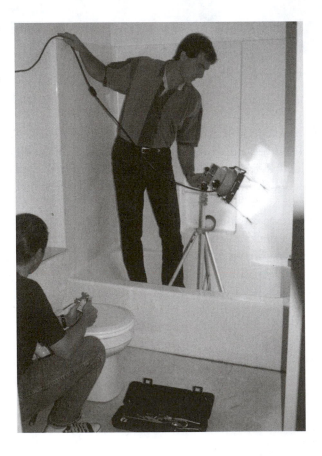

Review: Six Reasons Why Two Is Better Than One

1. Shooting with two cameras will help because the action will not have to be repeated. If shooting in a wide shot with one camera and a close-up with the other, you have the material sufficiently covered. This is called "covering your assets" (if you own the cameras).

2. It's more expensive to shoot with two cameras (the other operator and the cost of the camera), but you will shoot the material in half the time. This just about evens out the cost. Make sure each operator knows what you want covered and you will have an earlier wrap.

3. Both cameras should have an identical time code (jam sync it) to aid in editing. If the editor is looking for a close-up of the action, he or she can find it immediately on either camera's tape. If you get your editor mad at you . . .

4. Talent prefers doing the action once rather than repeating it. If shooting an installation, sometimes the action cannot be repeated. Installing a floor once is fine, but ripping it out and doing it again in a close-up will have you wearing pieces of the shattered floor.

5. With two cameras shooting, the video may be edited in numerous ways. The editor can cut to a close-up anytime and potentially all this footage would provide several versions of the completed program. In a DVD, you can finally get to use that multiple angle feature.

6. If cramped for space (two cameras take up more space than one), stick or clamp the lights in unusual places to keep them out of the way. This worked for us in the bathroom; just watch what you're clamping onto.

Making a Video Your Mom Would Be Proud of

We all want to please our mothers, don't we? Whenever I shoot a video, I'm always interested in what viewers will think when they first see it. Are the production values low and does it show? Is the concept easy to understand or does the viewer become lost early on? My desire is to have anyone who sits down (you can't watch standing up) to view my program be entertained, informed, enlightened, and not feel like the process had been a waste of time.

I recently shot a short video where all the action happened in the back-seat of a car. Not having used car mounts in almost 20 years, I had to learn the new technology and quickly become an expert in its use.

Having Your Car Mounted

Car mounts are like small furry pets: they need constant attention, they have to be adjusted constantly, and the second you turn your back on them, they fall off the car. On the other hand, car mounts are the only way to capture nifty footage while the car is moving.

The vision of director Robin Reck of Grand Trine Productions, "The Cry-ers" is an amusing piece about parents picking up their daughter from college and meeting her new boyfriend for the first time. Since most of the action takes place in the back seat of their 1992 Buick Century (What could happen? Their parents are there!), the camera, lighting, and sound equipment had to be hidden from view.

The car was mounted on (or driven onto) a U-haul trailer and towed behind a 10-cylinder U-haul truck. The driver of the U-haul would sit in the confined comfort of the heated cab and the rest of the crew was the hungry cargo in the back. Five actors would reside in the car atop the trailer as they were towed through the winding, secluded roads of a state park.

One of the first challenges was illuminating the actors inside the confines of the vehicle. Daylight would be the obvious choice in lighting because so much of it was pouring through the windows. The original vehicle was to be a 1966 Dodge, which would have been much larger and easier to light without the advent of tinted windows, but the sheer girth of the antique was too much for the car trailer.

The side and back windows were gelled with Rosco ND 9 gel and gaffer-taped to the car's trim. This cut the daylight filtering through the already tinted glass and softened the harsh blow by two stops. Two four-light Kino-Flos were rented and mounted to the car's hood (on the camera mount) and blasted through the windshield. Each fluorescent tube was daylight balanced (5,600 degrees Kelvin). This diffused, key light would cast a cool glow on Mom and Dad in the front seat but offer little on the kids in the back.

To solve that problem, four Micro-Flos were attached, two on the front dash and two on the backs of the seats, via the magic of gaffer tape. The daylight-balanced Micro-Flos added a sparkle to Mom and Dad's eyes and provided a little more light for the backseat occupants. Powered via dimmers, the Micro-Flos in the back were cranked to full, while the two in front were half-cocked.

Two digital cameras were employed to capture the action from two vantage points. They both performed flawlessly on 60-minute tape loads. A full-size Betacam or Digi-Beta were also designed to fit the mounts, but the director preferred the 60 minutes of our mini-DV.

All six lights and the video monitor were powered by a 24-volt marine battery tied into an inverter that changed our AC to the battery's DC. On breaks, the silent unit would be powered off to conserve energy. Weighting twice as much as my car, this hefty plastic enclosure could power everything except my space heater.

Two lavalieres were used to separate the front sound from the back. Taped on the dashboard and rear folding armrest respectively, these omni-directional units recorded the dialog in two channels. Unfortunately, the cameras did not have XLR inputs, so an XLR to a mini-adapter extended from the back and allowed perching for errant birds. The cabling, dimmers, power strip, XLR cables, microphone battery backs, and lights were all hidden but took up much needed space in the car. The sea of electronic spaghetti was ankle deep as the extension cords and XLR cables were fished through the back seat, into the trunk, up over the lid, taped to the side of the car, down onto the trailer, wrapped around the towing hitch, and plugged into the inverter.

The two camera mounts were also rented. I know I should have paid more attention to my erector set as a child because the camera mounts unpacked as a thousand two-inch-long pieces sprinkled with several nuts and bolts of varying lengths housed in two huge Anvil cases. The hostess tray (not my name for it) was assembled and clamped to the driver's door for camera position one, while the second camera was mounted on the hood between the lights and shot through the windshield on the hood mount. In order to prevent the hostess tray from marring the pristine finish of the Buick, the rental house included Staples mouse pads to put between the car mount and the paint. A mouse pad from Office Max might not have worked effectively. Make sure the mounts are attached securely to the car's body. This may sound like common sense, but the clamping grommets were the only thing keeping the cameras where they were. Close brushes with trees, massive potholes, or a stripped screw will cause your camera to fall from grace. Check and double-check every connection before using the mounts to hold cameras.

These are the only types of car mounts I'm familiar with. Without strapping someone to the side of the car, these mounts effectively hold any size camera where it needs to be. The hostess tray can easily be mounted by one proficient grip, while the hood mount, because of its size, needs someone on each side of the car to gently lower it into place. Once adapted to the car hood, shooting through the windshield demands extra attention.

Figure 5-9
An early version of a
car mount

Figure 5-9
An early version of a
car mount

If you need to shoot anything inside a moving vehicle, it's far safer to use a car mount. These units are designed for this purpose, are very inexpensive to rent (they won't kill your budget), and are the only way to effectively capture this footage. I've tried hanging from cars, leaning out of windows in moving vehicles, squatting in tiny trunks, or being gaffer-taped to the hood and nothing compares with a professional car mount. Figure 5-9 shows the extraction procedure when setting up car mounts.

Don't Try This at Home

My first attempt at makeshift car mounts in my youth was a 1979 film where I got an inflatable beach ball from Radio Shack and partially deflated it. This bean-bag-like gelatinous mass was securely taped to the hood of a 1975 Dodge Dart and the Super 8 film camera was taped to the beach ball and hood. The camera was pointed ahead and I slowly drove around bends, covered bridges, and country roads to capture subjective footage to be used behind the credits. This was a great way to break a $500 camera, imbed plastic shards in my radial tires, and have to dream up some excuse for my insurance company.

The images looked wonderful because the half-inflated beach ball and the car's shocks (before the advent of struts) made the film image float. Like an extremely cheap Steadicam, the camera wallowed and meandered along. Some people complained that the floating images made them queasy, but it was a horror film and that was my intent (right, just dumb luck).

I Can See Myself

Shooting through glass is difficult at best. This windshield shot presented several problems that had to be addressed individually. When shooting through any curved piece of glass, reflections will drive you mad (you should look through my glasses someday; I'm told I'm the only one that sees the yellow elephants). A Tiffen polarizing filter was attached to the camera and rotated until most of the reflections were eliminated (there's never a good Terminator around when you want one). However, the camera's image was still visible in the shot.

The camera was covered with a black Hefty trash bag, but the 60-mph winds caused the bag to rustle, flip, and fly into the shot. A yellow blanket was employed, but the jaundice reflection in the glass made everyone nauseous. Finally, the gray trunk liner was removed and taped over the camera. Figures 5-10, 5-11, and 5-12 show various aspects of the car mount.

Figure 5-10
Car mounts

Figure 5-11
Hiding the car mounts

Figure 5-12
The entire rig

This worked well when the sky was overcast, but the split second the sun peaked through we were scrambling for more blankets. Robin had a vibrant red casket liner that was taped over the cameras and windshield, and was extended with C-stand arms and sticks from a dead tree. The edges of the blanket were taped to the roof and tucked into the windows that couldn't be seen by the camera. The red sail inflated when the car moved and only blew off 17 times.

As the wind chill approached minus 11, I began to ice up. Both cameras had to be strapped (with strong bungee cords) to safeguard them falling.

Once on the trailer, the only entrance to the car was through the rear door on the driver's side. As the actors climbed over each other and were safely inside, the cameras were turned on and the operators jumped into the truck. Through a code system of knocks, one knock meant go, two meant stop, three meant speed up, four meant slow down, and 47 knocks meant stop at the nearest mini-mart because I want a coffee, a bagel, and someone has to use the rest room.

As the truck and trailer were moving, we crew people huddled in the open back of the truck and watched the action on a five-inch monitor. Like wounded refugees, we stayed in a close group to keep warm in the back of our metallic coffin to view what prevailed. Our clipboards could be used to ward off predators.

I learned many things from this incredible shoot. With a car in tow, you are unable to correct flaws immediately. You see something happening, but it involves much more than simply reaching over and adjusting that bag thingy. The towing vehicle must be stopped; you must jump out without tripping over the hitch, stop the cameras, and then address the particular problem.

Kino-Flos are the only way to travel because they output no heat and their illumination is flat, even, and constant (like a girl I once dated).

Mini-DV cameras are the best capturing devices because of their small size and full range of features. The key is to plan everything in as much detail as possible before the shoot. Make sure your car mounts are assembled properly and the cameras are attached securely, and guard your setup at all times, because someone kept leaving their empty mugs, half-eaten French fries, and catsup on our hostess tray without any tip! Figure 5-13 shows that eight can live as comfortably as six while living in a truck.

Figure 5-13
Refugees in a truck

Figure 5-13
Refugees in a truck

Review: Ten Things to Watch Out for When Using Car Mounts

1. Develop a knock system so you can communicate with the driver of the tow vehicle. In the metal cage you call home in the back, your hands quickly become numb from pounding on the frigid enclosure (your tongue sticks, too). Make sure the driver has the same knock code as the occupants in the back. Remember once you knock, you can't erase it.
2. Long before the actors arrive, assemble the car mounts. Although they aren't too complicated, we mounted each one upside down on our first attempt (just to see what it looked like) and wondered why the camera wouldn't attach easily. Never read the exploded diagram on the inside of the shipping case; that's only for novices.
3. Get everything out of the vehicle you may need before you exit. Entry is a challenge for all but the young and agile, because my hooked feet would constantly pull every cable and light with me as I bounded out. We did lose two make-up people somewhere in the car, but there's enough air in the trunk for at least three days.

4. When you're shooting outside in late fall, make sure the park hasn't winterized and closed all the bathrooms. Once winterized, a large bear, jumping up and down, and small unmarked bills will not make them reopen the facilities. I'm glad that woman 26 miles away didn't mind us using her bathroom.

5. Gaffer tape is your best friend on a shoot. However, the 32 rolls we used didn't adhere to anything because of the cold, velour surfaces everywhere and all the activity and sweat in the car. It makes a fun game at parties, what won't gaffer tape stick to (the dashboard, the car's metal . . .).

6. When shooting in the fall, the windows of the car should be closed to prevent the sounds of the trailer squeaking, the garbage bag rustling, wind whistling, and crew screaming from entering the confines of the vehicle. The drawback is that humans expel carbon dioxide, which fogs the windows. Either pretend you just installed a fog filter, make the actors stop breathing, or have them carry plenty of tissues (unused ones).

7. Everything shot on a moving car takes three times as long to do. Once the director is ready, the camera(s) must be started and you need to make sure tape is rolling. The operators have to climb back into the vehicle, the vehicle starts moving, action is called, the scene is played out, the director says cut, the tow vehicle must stop, the operators exit, and the cameras are then turned off. This routine must be repeated for each shot. Don't kid around by saying "Oops! Great rehearsal!" after a 17-minute take.

8. Everything you didn't plan on happening, will. The sun will come out, overexposing the shot, the sound blanket will fall into frame, the driver will run over a tree stump and cause two grips to roll out, the gaffer tape will give way at the wrong time, and the walkie-talkie in the car will lose its squelch. Communication is important. The director may talk to the actors through the walkie-talkie, and their reply is picked up through the microphone. Don't pretend you're a lounge singer at this time.

9. When driving under trees in a metal truck, the sound of branches hitting the top is no better than fingernails on a blackboard, especially when you just finished telling the "deranged man with the metal hook" story.

10. Be ready with responses for the gawkers that will happen by. When they see a car covered in ND gel, a red blanket, and gaffer tape on a trailer with two cameras and several lights attached being towed by a truck with nine people with clipboards in the back, they will still ask, "What are you doing?" When you tell them the truth, they will still respond with, "Seriously, what are you doing?"

Special Effects on a Low Budget

Special effects, if not overused, can add to any video production. Most of us don't have the money or access to Hollywood-style effects, but on a very limited budget you can create believable effects without the costs usually associated with them.

Honey, I Blew Up the Camera

I was asked (okay, paid) to shoot a visual montage showing young people fighting, breaking car windows, firing guns, and getting shot. The last scene in the montage featured a young man running toward the camera and then getting shot in the chest, complete with his shirt ripping open and blood spurting out.

The opening sequence was intended to shock the viewer. The last image would leave an impression (more so on the young man). The only trouble was the producer, Todd Taylor, wanted me to make the shot a reality.

I called a friend, Greg Ressitar, and asked if he had any information about making blood squibs (not on the backyard grill). A squib is a small explosive charge that can be detonated on command. They are usually placed on someone's body. He immediately sent me 14 pages from an Internet site that explained how to create safely exploding blood squibs. With some variation, I have manufactured glo-squibs. (They don't illuminate at night, but it sounds better than "Chuck's exploding blood packets.")

Before I divulge my secrets that everybody knows, let me emphasize that you exercise caution when making these devices. Although you are *not* using gunpowder, the mixture *is* flammable, so keep your little brother out of the room when you make it.

Alone in a darkened studio with Christopher Neu, the television production manager at Freedom Village and several other henchmen, we set out to create the perfect exploding squib "our way." (Frank Sinatra already did it *My Way*.)

The igniter is the element that makes the blood-filled sack blow up. Cut two five-inch pieces of insulated copper telephone wire .51 millimeters in diameter (24 AWG). This is interior telephone wire, so wait until you're off the phone before you rip it out of the wall. This will be called the lead wire.

Strip one inch of insulation from each end of both wires. Cross the wires at 90 degrees about 1/4-inch from the ends. Pinch the long ends of the wires in one hand. With your other hand, grab the short ends of the wires. Twist and pull the short ends of the wires until you have about a half-inch of wire twisted together in a tight rope-like spiral. You should have formed a small fork at the end of the twisted section. Make several of these units. Each time you ignite one it will vaporize. This is good therapy for the kids at night. It's safe, tedious, and more fun watching someone else do it and go buggy-eyed. All the materials needed for blood squibs are displayed in Figure 5-14.

Find someone with extremely good eyesight for the next step. Stretch one strand of copper wire from the left fork prong to the right, wrapping the wire around this prong five times. The fork's gap should be only 1/16 of an inch. This single strand of wire can be from an 18 AWG lamp cord. Just unplug the lamp before you cut into it. The single strand is very important. Initially, I used several (more than two) strands and the contraption will not ignite with more than one strand.

Now that you have the igniters, you need to ground (or mash potato) the powder. Purchase a one-pound jar of Pyrodex from Wal-Mart or any sporting goods store. Pyrodex is a muzzle-loading propellant that is smokeless and fireless if used correctly. Because the powder is granulated, you

Figure 5-14
Blood squibs

must crush it into a fine, soft powder. If you don't, the powder will ignite and smoke.

Pour a half inch (no more) into the cutoff finger of a disposable surgical glove and stick your igniter into the powder so the single strand lead is buried. I know that the finger isn't filled, but you need air space for the powder to ignite and burn effectively. Tie off the finger securely so it is blood- and airtight. If you don't, the powder will ignite and smoke. Don't smoke while you grind and don't have someone muscle-bound do it for you. Too much pressure and you'll be grinding on the moon.

May I Have the Condoms Please?

Mix up the blood and pour it in a condom. Yes, that's what I said. Put your exploding surgical glove finger into the condom and tie it. You now have an exploding blood squib. Attach a 25-foot length of lamp cord to the leads sticking out of the squib and attach the other end to a car battery (only when you're ready for it to explode). Twelve volts is needed to safely rupture this sticky, staining mess. My advice is to test at least five of these before they are attached to humans. If not, you will have fewer friends than you had before.

Hanging from a chain-link fence, we set off each test sample (I wasn't hanging from the fence; the squib was). Some burned too fast; some ex-

Figure 5-15
Condoms on a fence

panded the condom like a balloon and didn't break the blood sack. Experiment with more or less air space, finer ground powder, and so on. Never let your exploded condoms on a fence, as shown in Figure 5-15.

Once we were sure of ourselves (stop laughing), we attached our squib to a 4-inch-square piece of 18-gauge steel. This steel is what protected our victim, I mean talent, from the charge. Use gaffer tape to wrap the steel before you tape the squib to the plate. Leave a large section of the condom exposed in the front so the blood can rupture. Have your actor wear a T-shirt and tape this steel-plated squib to his chest over the T-shirt. If he is a weightlifter or swimmer, he may want his chest hair removed anyway, but let him do that on his own. Make sure the steel plate is secure, as shown in Figure 5-16.

The charge isn't strong enough to rip the fabric of the shirt so cut a small X in the shirt directly over the squib. Leave one strand of fabric still

Figure 5-16
Attaching the steel plate

attached so the shirt looks intact. The charge will break though the condom and the weakened fabric.

Using one camera to shoot the main action in a medium shot and another capturing it a close-up, we were ready. The poor soul would run toward the camera and the lead wires would touch the car battery for the explosion. With 25 feet of wire attached to the steel plate on his chest, running down his pant leg, and neatly coiled on the ground, your star doesn't have too far to run before he runs out of leash.

When the positive lead wire touched the battery terminal, a spark ignited our charge and sent eight inches of fire from our pre-cut hole. Clouds of noxious gray smoke poured from the opening, followed by a thick stream of spurting blood.

Although it looked great, few humans exude flames and smoke when shot (unless they are shot with a flare gun or bazooka). As the blood continued to ooze from his gapping wound, we realized that several more stunt shirts were needed to make a realistic effect. We had used too much power, too coarsely ground, and had too little air space for the powder to ignite, thus the smoke and fire.

Costing only a few pennies to create each squib, remember that this "blood," although tasty, stains (see Table 5-1). The best advice I can give is to practice and experiment until you have achieved the desired effect. Just use caution whenever working with powder and keep the blood squibs away from your eyes. If anything goes awry, you can write me in my backwoods hideout. With my days of pyrotechnics safely behind me, I enjoy tying fishing lures with my leftover supply of condoms. The realistic effect can be seen in Figure 5-17.

Table 5-1

Stage blood
formula

4 parts Karo® clear corn syrup

2 parts chocolate syrup. Hershey's works best

1 part red food coloring

1 part water (the stuff you drink)

1 drop blue food coloring per 59 cc ($^{1}/_{4}$ cup)

1 shredded tissue to simulate bits of body (gross factor)

Figure 5-17
The finished effect

Review: Ten Things to Remember when Creating Blood Squibs

1. While shopping, don't have a crazed look in your eye when you ask where the Pyrodex may be found. If you resemble a terrorist in any way, have a friend make the purchase.
2. When you purchase the igniter wire at Home Depot or Lowe's, know what you want before you ask for help. I stupidly had a print out that read, "How to Make Effective Igniters and Blood Packs" in clear view of the salesperson. He carefully cut the wire and ran from the isle screaming for the police.
3. Don't send your wife to the store to buy condoms, surgical gloves, and chocolate syrup in the same trip like I did. Whatever you do at home is your own business.
4. After depleting our six-pack of condoms during the testing, we went to the only country store in town to get more. Don't ask the attendant for "40 condoms that will break please." No one in that town dated that evening because we had exhausted the supply of condoms.

(continued)

(continued)

5. Use care when making the stage blood. Your crew will consume half of it before it enters the condom because it tastes good. Any spill will stain for days, and use tissues in the blood to resemble pieces of flesh exiting the wound. It looks disgusting, but it may get you an Emmy.

6. Clean up all your bloody, shredded condom remains after your testing. No one wants to find things like that lying around. No other jokes are necessary here; I think you have enough visual information.

7. When you return to your motel at night with bloody hands and clothing after making squibs all day, don't casually mention to your colleague in front of the desk clerk that you need to "shoot more kids tomorrow; we didn't get enough today." Instead, replace the word "shoot" with "videotape."

8. From the large onrush of volunteers you will get to "ignite," make sure they are aware you are using a mild explosive that will be strapped to their body. When they still line up, tell them that it stains their skin and will ruin their clothing. When the line is down to 50, tell them they will have to launder their own clothing; that should do it.

9. Experiment as much as possible with any deviation from your original formula. When a test works perfectly, you need to know exactly what you did correctly. Don't ask your friends to come and watch you "get blown up."

10. Don't let this be the start of bigger and better exploding things. Pyrotechnics is a professional's game and we shouldn't play without training or proper supervision. No one gives us a Get Out of Jail Free card in real life.

For more information, Pyrodex's web site is www.pyrodex.com or call 913-362-9455.

Table 5-2	(2) 5" pieces of insulated copper telephone wire .51mm (24 AWG)
Recipe for Exploding Squibs	(1) Single strand copper wire 1/2" in length (18 AWG)
	(1) One-pound jar of Pyrodex (P)
	(1) Disposable surgical glove
	(1) 25 foot length of lamp cord
	(1) 4 inch square piece of 18-gauge steel
	(1) Car battery

Documentaries

Getting Funding from Strange Places

The hardest part of making documentaries is getting the funding to do so. If you need to tell your story on videotape, who do you get to foot the bill? The easiest answer is the party interested in making the documentary. If they have no money or expect it to come from elsewhere, you now begin your search.

Shooting High Definition Without Getting Railroaded

The following shot was a documentary for Japanese television, but all of the money raised for the shoot was from the United States. Videos with an international flavor or appeal have an easier time obtaining funding because the final viewing market or demographic is larger. This Japanese program would be made available in this country on video and DVD (we share the same *National Television Standards Committee* [NTSC] standard).

Shooting high-definition video gives you the sharpness and clarity of 16mm film without the added expense of processing. I recently had my first exposure to high-definition video and now always want to use it.

Although physically not a child, I must think like a child, because I seem to get a lot of work working on children's shows. A little while ago I worked on the feature film *Thomas and the Magic Railroad* as a video assistant and when another *Thomas the Tank Engine* story came into town, I was available for that shoot.

Once a Japanese television crew came to Lancaster to shoot a half-hour video entitled, *Ponki-ki's: Go! Go! Thomas and Connie in America*. The video was shot in high definition to enable the scenic backgrounds to merge flawlessly with computer animation. The producer, Yukihiro Ito, and the production coordinator, Yuka Sakyo, from Office Kei in Manhattan contacted the Pennsylvania Film Commission in the hopes of securing the same locations used in the *Thomas* feature. The film commission contacted me and I arranged a five-day scouting trip through scenic Lancaster.

The 30-minute video featured an animated lead that would interact with Thomas and the various other characters in the story. The production was scheduled to occur when Thomas was in town. Thomas is a 1917 Porter steam locomotive that has been meticulously restored in Strasburg. Painted bright blue and red with a large smiling face on the front, Thomas attracts thousands of people wherever he goes. Our five-day shoot was scheduled to coincide with Thomas's arrival. If no advertisement for Thomas had been done, people still would have flocked to see him. There's something about a smiling train that brightens your day.

Four thousand people joined us the first day of the actual shoot. Because it was a Monday, the entire crowd consisted of young parents, grandparents, and little children. No child over the age of five was present. We had been infiltrated by the Class of 2016. As Thomas was parked on the tracks (the best place to park a train), parents would take turns sitting their children under Thomas's massive smiling face for a Kodak moment. A line of over 200 people formed during the train's 15-minute loading and unloading sequence. In that brief span of time, every parent had to carry their child over the chain-link fence while holding their camera, carrying case, diapers, soft wipes, and other personal effects in their teeth. If only children would realize the great risks their parents take for them.

The child would then be gently dropped on Thomas as the parent raced back to snap the picture. It was only then that the child would realize that he or she was sitting on the front of a steam locomotive and begin wailing with arms outstretched. Suddenly, Thomas didn't look so friendly. The only train with a smiling face can be seen in Figure 6-1.

Figure 6-1
Thomas the Tank
Engine

As the parent tried to calm the hysterical child, the next occupant in line would begin psyching his or her child up for the picture. It was during this photo opportunity that the crew and I had our only chance to videotape Thomas with Cardboard Connie. I was given the dubious honor of controlling the crowd while the camera crew got their shot.

Just put yourself in my place for a moment. You are a parent with two-point-three screaming children who have been nagging you to see Thomas for the last month. You had to take off work because it was a weekday and you want your child to remember this day forever. You also have been standing in line for several hours patiently waiting your turn. When it finally arrives, some bearded joker you never saw before asks you if you would wait a few minutes until the video crew finishes their shot.

The crew didn't stand in line all day, the director never nagged, and Cardboard Connie wasn't going to start crying when she saw Thomas for the first time. When the parents were done hitting me and their children had calmed down, life was once again pleasant.

The *director of photography* (DP) was brilliant in his shot composition. He knew exactly the height, distance, and angle in which to set up the camera. He merely pointed and his assistant had the camera leveled and ready to go.

The crew spoke very little English and that worried me at first. Both the A and B crews were directly from Fuji TV in Tokyo. But I guess video is an international language. Not once did I misinterpret what they wanted or needed for a shot. Video professionals are the same the world over.

As the director, Atsuki Yamazaki, watched from behind the DP, the first assistant would hold up a cardboard drawing of the computer-animated character (Cardboard Connie). This cardboard object was the same size and shape as the computer image and was used for size perspective. In post-production, that cardboard would be removed and the character inserted.

This young man carried his cardboard person with him throughout the day. He would be seen sitting on the train with the cutout next to him, at lunch with the character beside him, and on the way home in the van with it folded by his side.

The train filled with passengers prepared to leave and the video crew shot as the locomotive belched white steam and a fine mist of unburned coal. With the large number of tourists, the camera was always attended. Although outnumbered 2,000 to 1 by camcorders, the Sony 700 high-definition camera attracted little or no attention. Of course, I was fixated by the sleek lines and graceful curves of the electronic marvel, but I don't get out too much.

After controlling the crowds for quite awhile, I had a great idea. If parents were willing to wait in line for several hours to have their kids get pictures with Thomas, they would love the opportunity to have their pictures taken standing next to a state-of-the-art, high-definition camera.

I formed my own line and announced that if anyone wanted their pictures taken beside a $110,000 high-definition camera instead of Thomas to stand behind me. All I received were blank stares. Thomas was always around, but their chance to be with a high-definition camera may not happen again soon. I guess I'll never understand people.

While on the train (this time as actual passengers), the A crew adjusted for the backlit conditions and recorded the passengers as the scenery passed by. The teak and mahogany of the train's interior was six stops below the exterior of the train. With white foam core, soft white daylight was bounced into the passenger compartment, making the dark seats more visible. This is the only time any of the passengers even noticed this camera. It was probably because they noticed the white foam core moving around.

Once again, I've never seen a crew who knew their cameras as well as this crew. While speaking softly to each other in their native tongue, the assistants knew exactly what the DP wanted and then made it happen. The sign of a great crew is one that anticipates what the DP wants before he or she asks for it. Either that or hire psychic crew people (not psychotic, there's a difference). Figure 6-2 shows the high-definition crew at work.

Figure 6-2
High definition at work

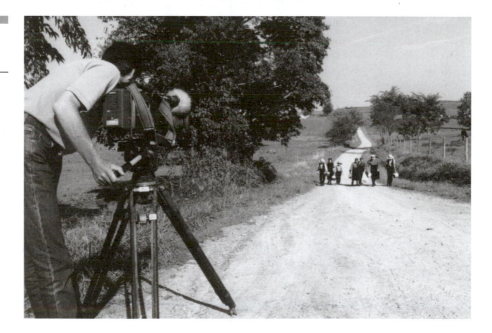

Please Don't Shoot, We're Amish

The Amish of Lancaster, Pennsylvania, with their origins in sixteenth-century Europe haven't changed that much over time. They still use the horse and buggy for transportation; have no need for telephones, electricity, or zippers; and believe that capturing their images on video or film is making a graven image. What better way to tell their story than with a multimedia, high-tech, theatrical experience?

"The Amish Experience" is a three-story F/X theatre located in Bird-in-Hand, Pennsylvania, that tells the story of a young Amish man's struggle with his heritage. I was involved in the creation of a 28-minute video (the centerpiece of the Experience) called *Jacob's Choice*.

The funding for this video was from "angel" investors. These angels don't have wings but are very nice. Through investments, these people invest large sums of money to make videos a reality. Where do you find these types of people? The best answer is any place you can.

Create a prospectus of your documentary. The prospectus is the proposal that potential investors read and determine if your project is worth them investing money. Obviously, they would like to make a profit, but serious investors know it takes quite a bit of time to recoup video money.

I'll go into more detail in Chapter 9, "Features," but the same principle applies for any video. Advertise a screening of your previous work or a short piece of what you propose to shoot. This viewing will show investors you are serious and can get the job done. Through this screening, the following project found all its money.

Armed with a camera and an 11-day shooting schedule, we spent our time wisely on the fields and farms in Lancaster. The actual costumes used in the movie *Witness* were rented from Paramount Studios for this project. A coat bearing the initials of Harrison Ford was also used.

Although non-Amish actors were used as the principles in the video, actual Amish were extras. When shooting the wide expanses of Lancaster County's countryside, curious children would somehow find their way in front of the lens. Adults were a different matter because they would lower their heads when they caught sight of our camera. After miles of B-roll had been recorded and all the principle shooting had been completed (in actual Amish cemeteries, barns, and farms), we recorded the studio material.

Known to many as chroma key boy, I personally hand-painted a 20-foot stretch of canvas with chroma key blue paint. If I had inadvertently walked onto a weather set after a hard day of painting, I would have had maps and storm fronts where they shouldn't be. The key thing to remember (pardon the pun) is to use a roller if painting. The blue finish must be uniform and even (just as its lighting.)

One of the most difficult scenes to shoot was the ghost sequence. Actors dressed in sixteenth-century costumes stood in front of my evenly lit chroma key blue screen. Fire was also recorded against the same blue screen to be used later. The blue background would be keyed out and replaced with black. Ghosts seem to like black better, so why fight a good thing?

This sequence was edited at Varitel in Los Angeles and composited onto a D2 master. Using Discrete Logic's Flame, the actors would magically appear in waves and then soon disappear in a black void. The special effects designer, David M. Blum, also had the distinction of helping create the original *Star Trek* transporter beaming effect. My blue paint was seen by some important people.

Using a beam splitter back in the actual theatre, the ghosts could come and go as they pleased. A beam splitter is a partially silvered or mirror piece of glass that allows you to look through it, but also reflects a portion back to you. Actually, we made them do what we wanted; ghosts have no union. Rochester Insulated Glass Company manufactured a single piece of glass weighting 800 pounds. This was the largest glass ever produced by Rochester. Using six people, the glass (with one partially silvered surface) was mounted on a 45-degree angle on a stage 5 feet deep by 10 feet wide.

A *liquid crystal display* (LCD) projector mounted 20 feet above in the attic of the theatre (in an ingenious wooden and ventilated enclosure designed by one of the best blue screen painters in the business) would project its image onto the glass and reflect part of it back into the void created in this blackened room. A thin mesh stretched across the front surface of the room acted as a viewing screen. The ghost images would then be contained in this enclosed room.

Instead of using normal projection screens, the designers chose to use actual wooden barn boards. Each plank was painted white (for better reflectance), gapped an inch apart, and hung to create the screen. Thin muslin was stretched behind the wooden boards for its entire length.

The end of the last shooting day involved us "Amish chasing." When we would see a horse and buggy or an Amish individual pushing his scooter along, we would race ahead of them in the van and stop at the top of the next hill. The crew would disembark, set up the tripod and camera, and record the soul riding toward the camera. When they raced past, head lowered, the crew would grab their camera and run along behind them as the talent made their way up the steep hill.

Figure 6-3
Amish and cameras

After recording the same Amish teenager several times on his scooter, he pulled up alongside our van and asked, "What are you doing?" in a heavy Dutch accent. The first thing that came to my mind was "We're shooting a movie." "I thought as much," he replied and was on his way. A prime example of how much Amish like cameras can be seen in Figure 6-3.

This game of cat and mouse kept us one step ahead of the Amish and also afforded us some excellent footage. I have been trying for years to get this type of footage to keep as stock shots of the area but had been unsuccessful.

Now, with the last remnants of blue paint slowly fading off my skin, I can fondly look back and remember the summer when I was Amish (except for the mustache, beliefs, and narrow choice in clothing colors).

Review: Ten Things to Do when Working with the Amish

1. Don't tell them you like the color of their wardrobe (they only wear black).
2. Don't let them look into the camera. Most have never seen video before and will never leave.
3. Never ask them if you are doing it the right way. You aren't and you couldn't possibly.
4. Never say your car is faster than their buggy; it may use less straw.

5. They are perfectionists and will do something until it's done correctly. This is something we may need to learn.

6. Never ask if they will help set up your light stands. They have worked in the fields and are extremely strong. Once they tighten a C-clamp, you will never open it again.

7. Terms like "action," "cut," and "play that back" are foreign to them. Explain what you want in simple terms that aren't film or video related.

8. When they ask you when you want to start the next day, don't say, "First thing in the morning." That means 4:00 A.M.

9. Try to be sly when videotaping extras. They don't like being taped so if they see the camera, you won't see their faces again.

10. If they ask if their children can watch, be ready for crowd control.

Who Pays for All This?

Somebody has to pay for your documentary and too many times it's been me. Many times this "funding" comes from the sale of the documentary itself on video or DVD. Ken Burns received grants from Public Television for *The Civil War*, *Baseball*, and *Jazz*. But I needed to sell *Wanda, My Life as a Mule* on VHS before I saw any money.

Working with Cooked Vegetables

I enjoy vegetables as much as the next person, but when they're out in the summer sun all day they can become a might cooked. But that wasn't the case with Larry and Bob, the cucumber and tomato duo from the popular video series *VeggieTales*.

The nine-foot-tall green cucumber, Larry, and his pal the four-foot-tall red tomato, Bob, were mobbed everywhere they went. These two vegetables and the rest of the cornucopia of mixed vegetables in *VeggieTales* try to instill wholesome values in kids through their message and various encounters. These vegetables are role models for kids like the *Davey and Goliath* and the *Gumby* series were of the late sixties. *Barney* would be a similar present-day comparison.

It was my extreme pleasure to spend three days with Larry and Bob shooting a documentary. I was hired to be one of five video crews to make a two-hour documentary of Creation '99. This four-day festival featured over 30 big music groups, solo artists, and speakers that performed outdoors on stage before 110,000 people. As part of the documentary, I would follow Larry and Bob throughout the site and record their various encounters with the people.

Armed with a camera and a soundman, Tom Landis, we followed the dynamic vegetable duo through more than 200 acres of people, food, and attractions. When completed, the video would be sold at $14.95 and made available to patrons at Creation '00. By Creation '02, over 60,000 of these cassettes have been sold and the producer is now making a profit. Sometimes it takes several years for a profit to be realized. Almost no low-budget production like this one will start making money on its own.

Without bursting any bubble for the younger readers, Larry and Bob aren't really living vegetables. Both are humans that don a green or red costume. The two individuals wear a harness that supports the inflated costume. Powered by Nicad NP-1 look-alike batteries, air is sucked in through one hole and a fan inflates the costume to a cool and comfortable temperature. Air is then exhausted through another outlet (just like me).

Used as a running gag throughout the documentary, Larry and Bob were attending the festival and wanted to blend in with the rest of the patrons. Even though celebrities in their own right, they preferred to do what everyone else was doing. The viewer would find them turning up every 10 minutes or so throughout the video and see what mischief they were getting into.

Because the festival is four days in length, most patrons have to camp at the site. Hotels are available within a 60-minute drive, but over 100,000 attendees would still need other lodgings. Therefore, Larry and Bob had to camp out (some hotels are prejudiced against vegetables).

Their first scene in the video involved them waiting in the shower line. The festival has four permanent shower facilities throughout the grounds. Unless people bring their showers with them, this is the only way to get clean. The camera established a wide shot of the shower facilities with a long line of women going in the right and men going in the left. As I panned the camera to the left, men would be waiting patiently for their turn in the building. After the camera panned past eight people, it would continue panning until the end of the line, right past Larry and Bob. The director, Tim Landis, wanted the viewer to get a brief glimpse of the green and red veggies without calling attention to them. The viewer would see them only for an instant.

Other men in the shower line were interviewed and asked how long they had been waiting for their turn inside. As they answered the question, Larry would wander into the frame, towel wrapped around his waist, and look at his watch. Because of Bob's girth, three towels were pinned together to wrap around his 120-inch waist. The towels were held to his cloth body with gaffer tape. This subtle approach added to the humor because the large green and rotund red people were just trying to blend into the crowd.

Because the veggies had lots of time on their hands, they decided it was best to find work between the stage artist's performances. They went to the staff office and applied for jobs. Their position was directing traffic in campground K. Figures 6-4 and 6-5 show behind-the-scenes occurrences with my favorite vegetables.

Figure 6-4
Larry and Bob
directing traffic

Figure 6-5
Videotaping the
event from a golf cart

Campground K is three miles from the main stage area. It is the farthest campground from anything. The only way to get there is to drive or hike. Since Larry had a hard time fitting in an automobile, he and his sidekick were transported to Campsite K in an open golf cart.

En route to the location, both vegetables were mobbed by their fans. No one had been told that we were making a documentary (purposely), so the arrival of a golf cart with two vegetables and two other carts filled with crew got people's attention. While driving to Campground K, we videotaped several traveling sequences.

I mounted the camera close to the ground and had the Veggie-laden cart drive toward the lens. Because of the low angle and a large hill, the viewer would first see waves of heat rising from the surface of the dirt road. Within moments, a thin green head and two white eyes appeared over the top of the hill. Shooting through a 120mm lens, the background was compressed and the veggies seemed to take forever getting to the camera. As clouds of dust swirled behind the moving cart, traveling music would be played. A Sony Mini-DV mounted on a Digital Steadicam was used to record the veggies' perspective shots as well as 180-degree views of the traveling cart.

The two salad ingredients were driven to the center of an open field and left there with red flags. They were told to direct the cars to their campsites. I set my camera up in the back of my golf cart and the Steadicam recorded the closer action. Every resident of the camp had crowded around the perimeter of our shoot. Parents kept their children on the sidelines as we set up the shot.

Seemingly left alone to rot out in the field, the two fruit wannabes stood there waving flags. With temperatures in the low nineties, they began to wilt. As time progressed, they slowly crumpled, fell over, and fainted from the heat. The onlookers laughed as the baked vegetables tried to stand up for a second take. Because of the heat and the costume, the tomato needed assistance (all this was staged; no vegetables were hurt).

As part of the scenario, their employer returned in the early evening to see how they were doing. Seeing their collapsed bodies out in the field, he helped them into the golf cart and drove away. To simulate the passing of time, the camera was underexposed four stops and shot through a booster blue gel.

Deciding that field work wasn't working out, they were given camera operating jobs on the main stage. Larry was asked to run the jimmy jib and Bob was his grip. Once again, no one in the audience was aware that we were making a documentary. After coordinating it with the onstage performer, Larry and Bob were to walk out on stage and relieve the first shift camera crew. As the performer continued talking to the 40,000-member

Figure 6-6
Vegetables on stage

audience, Larry walked over and took the jimmy jib from its operator, Hank. Not knowing he was being relieved by a cucumber, the operator fought to keep his position. The last known photo of vegetables on stage is Figure 6-6.

Never having worked a jib before, Larry swung the live camera around and knocked Hank to his knees. The director, Don Barley, switched the actual image from Larry's jib to play on the 20-foot Jumbotron monitor. As the jib swung back and forth, nearing missing Hank's head, the audience roared at the images on the Jumbotron. Being a professional, the performer continued with his act as if he had no clue what was going on behind him on stage. Realizing that this wasn't working out either, both members of the tossed salad were removed from stage.

The next shot featured a chef preparing food at his counter in the food court. The vegetable's agent walked up to the gentleman chopping vegetables and asked him if he could use two hard workers who said they "wanted to work with vegetables." Without looking up from his chopping, the chef agreed that he could use some help for the dinner rush.

Larry and Bob entered from screen left and noticed the chef was slicing a tomato. Larry and a horrified Bob looked at each other and slowly snuck away, backwards. The chef looks up, shrugs, and continues his preparations.

Throughout the rest of the documentary, the two veggies are seen in various situations, each time doing a job that doesn't suit their skills. Only able to spend 30 minutes inside the costumes because of the heat, the actors would disrobe until we arrived at the location. There, under cover, they would dress and begin their routine.

I learned quite a bit from this vitamin-A-enriching experience. The "VeggieTales" characters are extremely popular with kids. Once we arrived on the location, we would lose 10 minutes of our narrow 30-minute window waiting for autograph and photo seekers to complete their tasks. Even though we had crew members out of frame with signs that told of our videotaping, the excited kids would sneak through to hug one of the vegetables. The kids would remember this for years, so we let them have their fun before the camera rolled. When shooting any documentary, you have to allow what happens naturally, shoot it, or wait until you have an opportunity to tape. These characters are popular and therefore their fans needed to be taped as part of their normal day.

When working with celebrities, crowd control is a definite. Without people-wranglers, the vegetables would have been whisked off by the crowd. It's also extremely important to set up quickly; you have to be ready when the vegetables are (how often can you say that in real life?). Most times, you can't do a second take; it's much like live television. You have to be prepared to record what happens correctly the first time. This strengthened my skills as a camera operator. I have to be positioned in a comfortable spot, check focus and exposure, and make sure I can pan, tilt, and zoom if necessary. Not knowing exactly what the talent would be doing in a specific shot keeps me humble.

A documentary camera operator needs to know his or her equipment intimately. Fumbling for the exposure takes up valuable time you may not have. If using unfamiliar equipment on a shoot, practice as much as you need beforehand to learn the unit inside and out. This not only avoids embarrassment, but it will help you immensely.

Wandering Aimlessly

The director also wanted us to wander through over 200 acres of campsites to capture what was happening in the various tents. We met people who lived in tepees, inflatable tents, motor homes, and canvas tarps hung over a wire.

As Tom (the soundman) and I wandered up unannounced to one of the lodgers, we'd ask if we could videotape them. If they agreed, we'd shoot and have them sign the release form later. No matter how low the budget, have every breathing individual who appears on camera sign a release. This allows you the right to use their image.

You Can't Please People Some of the Time

I shot a commercial where we asked shoppers why they liked a particular store they frequented. These people would reply with the first unscripted thing that came to their mind, often not making grammatical sense or being very humorous. We recorded a young woman who made a silly statement on camera and the store owner loved it. Her "stupid" remark was part of the donut and it appeared 12 times a day over the course of 6 months.

She called me at the TV station and said she wanted to be removed from the commercial because "my friends are making fun of me and I'm sick of hearing myself say something so stupid." She had signed a release form and was given a $25 gift certificate to the store (money doesn't always have to exchange hands, but if the budget permits it's a nice gesture).

Since we had a signed piece of paper allowing us to use her likeness and voice (we did), we legally did not have to remove her. It would cost us money to pull the old spot, find new footage of the same length, reedit, and reduplicate. If she had not signed a release, she legally could have the video pulled. We agreed it would be good PR to remove her and eat the expense (by remove, I didn't mean having her "whacked," but she was whacked editorially).

Sticking a camera in someone's face, even if asked, is an invasion of their privacy. They may be doing what they normally do, but you are shooting it and might be showing it to others.

My point in bringing all this up is that I wasn't totally familiar with my camera, and in a split second I had to travel from inside a tent to outdoors, adjust exposure, frame from a 1 shot to a 12 shot, and had no idea where my victim might be moving to. I won't go into the heat, humidity, my searing skin, fogging glasses, or hairy legs I had to deal with.

Everything must be on manual, and everything must be controlled (you ride the visual levels). Even the soundman had to point the microphone at whoever talked and change audio levels on the fly. This demands a good operator. It's great to be spontaneous, but you have no clue what's going to happen next. You asked them if you could shoot them and they can do what they darn well please. Some people, seeing the camera (or maybe just me), just said no. If that's what they say, you must respect their wishes.

In film school, my instructor Jerry Holway said when shooting a documentary "keep your left hand on the lens so your finger can change the exposure, roll the focus, and zoom at the same time if needed." Of course, this was with a film camera before the advent of power zooms and everything relating to image capture was on the lens. Now with video, the exposure control could be in the back or side of the camera, the zoom on top where the tape resides, and the focus a tiny wheel under the camera.

The same principle applies wherever the features may reside. The key is to know where they are and be able to adjust each of them for every shot. This takes dexterity, timing, and mostly luck (and skill).

One last bit of advice when shooting in a documentary is to establish every shot first, zoom in, get a focus and exposure (when you can), and set up your shot. This has to happen in a millisecond and everything you adjusted will change—be ready for it. It's best to leave the camera running during all this. Stopping and starting the camera to save tape only slows you down and the tiny bit of wasted tape in your setup process is time saved. Sometimes I'd be zooming in to get my focus when the person was doing something that was unrepeatable. It may have been a crappy shot, but I got it on tape. If I had stopped the camera to do my adjustments, no one would have seen it (including the tape).

Good Thing I Was Ready

I was shooting a day-in-the-life documentary for a law firm when I'm glad I kept the camera rolling. The woman was just getting out of the shower (don't ask; just go with me) when her five-year-old son opened the window of his bedroom and let his two-year-old sister walk out onto the roof.

We heard his screams from the bathroom and his mother ran from the shower to retrieve her daughter from the ledge. Still taping, I followed behind as she hung out the window, scooped up her daughter, and possibly saved her life. This love, compassion, and willingness to do "whatever it takes" helped the woman win her case.

If I had stopped the camera, this split-second emergency would have been lost. She and I had no idea this was going to happen, but we had to "expect the unexpected" and just be prepared. Of course, this footage was unbelievably difficult in editing because I had to mask out her "naughty bits" as she was flying through the house (a moving mask is not easy when everything is jumping around).

Stick to your word when shooting a documentary. If you tell someone you won't do or show something, do what you say. I told this woman that I would not show any parts of her that would embarrass her. No one else had to see those features and they would have no impact on the case. When shooting, I would frame out things that shouldn't be shown and since I never expected the birthday suit streak, I had to keep my promise and did much more masking in editing.

More Ideas for the Frugal

When having no money is a way of life, you still need to acquire things without having to spend too much money. If you happen to be shooting with a studio configuration camera at a concert, sporting event, or some rich person's wedding, how do you keep the glare from making the monitor image disappear? The answer is a viewfinder extender, which will keep the blazing sun out of the pristine monitor. Don Barley of RCTV Productions came up with a strange idea (I'll talk about that later).

Figure 6-7 shows the art-deco version of the cardboard viewfinder extender. Depending on the shooting arena, western and traditional (welcome

Figure 6-7
The RCTV art-deco viewfinder extender

race fans) versions are available. But what happens when it's 100 degrees in the shade and your face is stuck in this cardboard enclosure? This is where a fan (a race fan) is helpful. Small scroll cage fans or those that cool computer *central processing units* (CPUs) can be purchased for a few dollars. Attach the 12-volt fan leads to the camera's battery terminals (also 12 volts) and your hair will be blowing in the wind (better than the answer blowing in the wind). Try to keep the fan off to the side of the monitor because the magnet in the back will cause the *cathode ray tube* (CRT) image to pulsate.

The most important thing to remember is to decorate any makeshift device. This is done for two reasons: one, people will see it and it can't look cheap and crappy, and two, no one will throw it away if it looks pretty.

A disposable solution to covering your camera from rain may be solved by using trash bags and duct tape. It may not be pretty, but you can also use it as a sleeping bag (before you throw it away) if your shift on camera becomes too long.

Back to the Vegetables

In all honesty, I was fortunate most of the time capturing the action as it occurred spontaneously. Other times I needed to reshoot.

One last note: Usually, people stop what they're doing when they see a camera crew. I've gotten used to attracting attention with the camera. I don't mind it; it goes with the territory. On this shoot, people couldn't care less about the video crew. Not one person ran up to our golf cart. The veggie cart was a kid magnet. I was upstaged by two oversized, sun-dried vegetables. A humbling experience, but I don't have seeds.

Review: Twelve Things to Document when Shooting a Documentary

1. If your documentary is going to be humorous, have the lead character(s) in reoccurring roles or situations. Some people get very distracted when a video is nonlinear. But if Heckle and Jeckle keep turning up in various situations, the viewer will identify with them or feel continuity with the characters. You all know the trouble Heckle and Jeckle get into.

2. If you're using humor, be subtle. Don't hit the audience over the head with it. If the joke is supposed to be hilarious and it flops, everybody looks bad.

   More outlandish visual humor tends to offend some people. You can't please all the people all the time, but less is usually more in this case.

   The subtle approach lets the humor happen. If it goes over the viewer's head, it will be funnier if they think about it later and laugh. This is an example of a delayed joke. When it finally hits, they will remember it much longer.

   Pan past the characters so the audience may get a glimpse of them, but don't dwell on it. Sight gags carry more weight because their visual aspect will leave the audience laughing.

3. If doing a documentary on a famous person, follow them around and let what naturally happens occur. They will be approached by autograph seekers and someone who disapproves of them will tell them so loudly. Notice the reactions they get when in public places. This is part of their average day and should also be a part of the video (unless you're shooting one of the Osbornes; the Osmonds are a different story).

4. Put your lead characters in real-life situations; this can also add to the humor. The characters will just do what they normally do, but if they're doing something they aren't familiar with: humor! You should have seen me when I got married (I meant by being in a situation I wasn't familiar with, not anything else).

5. To simulate the passing of time, don't use a clock. Instead, accelerate the action of the video to show hours flying by, underexpose four stops to pretend it's night (and use a blue gel over the lens), tell the viewer via CG graphics on the screen (two hours later), or use a transition in editing as a last resort (the *Star Wars* wipe). Of course, you can tell the whole story in real time, but only Andy Warhol could get away with that (he doesn't anymore).

6. Don't force things to happen because you want to videotape them. Most times if you stage something, it never looks realistic. If letting something happen naturally isn't in the script, maybe it's important enough to be added. Like when that swarm of locusts engulfed our car, making us late to the premiere.

7. Be ready to set up your equipment quickly on a moment's notice; pretend you're an important doctor on call. You never know what is going to happen and you should be ready to shoot it (with the camera).

8. Know all your equipment inside and out, and know what it will do. If using new equipment, practice before the big day. I've been embarrassed more than once not being familiar with my camera in an important event (if I had been wearing pants, that might have helped too).

(continued)

(continued)

9. Know where the focus, zoom, exposure, and all the other features are and keep spare fingers on them (preferably yours). If you can find these features in the dark, you can use them when needed. But if you're really in the dark, the camera won't do you too much good.

10. If a person in any form appears on tape (even if you edit him or her out later), have the person sign a release form. As the name implies, it releases you from danger. If Will Robinson had signed one in *Lost in Space*, the robot wouldn't have had to constantly remind him.

11. When shooting, don't stop your camera. A documentary can't always be scripted and you may see something spontaneous that should be taped. If the camera is turned off, that makes difficult shooting. Tape is cheap enough to roll, but do turn it off when you go to bed.

12. If possible, frame out things that you know will not be used in the final video. All of these reality shows use the framed-out feature often so you never really see too much (that's why you keep the other eye open).

Choosing the Right Market for Your Documentary

Once your documentary is completed, how do you get it in the hands of the greedy public? If no one ever sees your project, it was only a practice exercise, probably an expensive one. Marketing is where you can really get your video to move. Like I mentioned, the Creation '99 documentary was sold at the next event, advertised in brochures, and several other avenues.

Never limit your marketing to just one place. If you shoot a music video, everyone in the band gets a copy, but how do you market it after MTV or VH-1 has seen it? Try selling the tapes or discs at concerts; give them as promotional pieces to radio stations, local book or video stores, and venues where the band usually plays.

The key thing is getting your documentary *seen*. This is the best form of advertising. Even if you do a documentary on something that sounds really interesting like "Raising the *Titanic*," people need to see some of it to help the sales process.

Poor Little Me

My first documentary way back in 1981 was made for no money. Called *Beach Movie*, I traveled to a sunny beach during my vacation and filmed people doing what they do best, being themselves. I just wanted to show on 16mm color and black and white film the sorted types that frequent the beach in droves. I shot large people, little kids playing, the sexy and the sleazy, all over the course of a weekend.

I metered every shot; it was always F11. I wound the camera, lay in the sand, and waited with my telephoto lens. When the color film was processed, they had an accident in the lab and two-thirds of it was fogged (the film, not the lab). The black and white footage worked well, but the color looked horrible. What was I going to do?

I decided to change the premise of the documentary and make it look like excerpts of people's home movies. Using little snippets here and there, I created a three-and-a-half-minute epic cut to music. By changing the premise of the film after the project had been shot, I saved its life.

In order to make some money to pay for the processing and printing of the film, I had it shown in local theatres. My film was 16mm and most theatres are 35mm, but in college at that time 16mm projectors were used much more. The same thing should be true with your low-budget documentary: Get it shown in front of large groups.

If your video appeals to environmental groups, show it to them. Find the audience for your classic and get them to see it any way possible (legally). Don't show your video on the process of mink coat production to an animal rights group. Once your video is shown (whether a paid showing or not), have copies available afterwards for sale. The more copies floating around, the more money you'll make.

Take Two

I documented the process of making apple cider from apples growing on the tree all the way to selling it in the plastic gallon containers. This video was funded by my local PBS station. They had seen and aired *Beach Movie* and asked if I would do a four-minute video on apple cider. I had to find camera equipment, editing facilities, and stock on my own; they would pay me each time it aired on their station.

I had a project, I had funding, once it was completed, and it had a market in which it would be shown. If you ever get in a situation like this, ask the funding source who has the rights to the completed piece. In my case, they would own a copy to air when they wished (and I would get paid each time) and I could do what I desired with the other copies.

I took copy number two and sold it to HBO. The same day I mailed copy three to Cinemax (Max). They both paid me each time the project aired. With these two additional sales, I paid off my college loan and bought a new car (things were cheaper back then). This no-budget video paid for itself the next nine years it aired on pay television.

But that's not the end of the story. Whatever happened to the copy that the PBS station owned? As I said, I would be paid (I believe it was $100) each time it aired, so I needed to get it on the air many times. It happens I was dating the traffic coordinator at the PBS station at the time. Her job was to schedule what aired and when. My little documentary, running only four minutes on noncommercial television, could be plugged into a lot of holes. That's exactly what she did. Every time there was an opening of four minutes or longer, she would schedule my video, it would air, and I would get paid.

She did this so many times I had to thank her, so I married her. I had to. Now when people say two people had to get married, I know what they mean. Nineteen years later we are still together (my wife and I), and the video has stopped airing (shortly after we got married and she left the station). My wife has been costing me money ever since (now that she doesn't push my videos).

I'm not saying that you should marry the traffic coordinator to get your videos aired, but find out who has that position and see what they will do. You will never know until you ask.

Who Me?

Now that you know some avenues for presenting your finished work, what other areas can you tap into to get cash for what you've done (not the stolen jewels, the documentary)? Is there anything special about what you shot or how you shot it that might appeal to others? It sounds to me like it's time for another story.

Shooting video footage from a helicopter is fun; don't let anyone tell you otherwise. I've been told that it's the most fun you can have . . . legally.

For that same documentary of Creation '99, I was asked to shoot aerial footage to open the video and be used throughout. Situated on over 200 acres in Mount Union, Pennsylvania, Creation is surrounded by, well, nothing. Bordered by forest and farmland on all sides, Tim Landis had a vision for the perfect opening shot.

The camera would shoot (from the air) miles and miles of trees and open farmlands. The shot's perspective would be just a few feet above the tops of the trees. As the camera flies over this breathtaking vista, a mountain top would loom in the distance. The brave soul in the aerial device would then slowly lift over the tops of the trees to reveal the valley below. Of course, there's nothing unusual about a valley appearing after a forest. The only difference was that this valley contained a 30-foot-high stage, a 20-foot Jumbotron monitor, and 110,000 people. That's the part that was different. Out of a sea of green, a massive stage and hordes of people would come into view. To add a slight realism to the shot, the producer envisioned the camera moving from side to side while shooting, like the memorable scene in *Apocalypse Now* without the Napalm. Much to my good fortune, the best camera to capture such an event was the one I was holding in my lap.

Stabilization is also a factor when flying through time and space. Hollywood has used stabilizer-mounted units for ages (Tyler mounts). We didn't have access to one and no one wanted to pay to rent one. (Remember, this is a book on no-budget video production. In no-budget, you hang out of the helicopter and shoot.) A Steadicam is also an excellent choice for through-the-air stability. Our Steadicams were built for the mini-DV units and wouldn't support the weight of a Betacam.

The 9mm wide-angle setting of our Canon zoom lens was wide, but I needed a wider field of view to reduce camera shake. The mini-DV palmcorders excel in this area because of their image stabilization. I borrowed a Century Precision Optics .8X wide-angle converter from our jimmy jib operator. This three-pound piece of glass gave me a wider perspective and would help reduce the jitter in the air. I now had a 29-pound camera that was nose heavy. At least the extra weight up front would help make it easier to point down. If I fell out of the helicopter, the paramedics wouldn't have to look too hard to find the camera; it would be buried 10 inches in the ground.

The next problem to solve was how to hold the camera. Should I hold the camera on my lap and frame it through the viewfinder or do I hold it on my shoulders and let my body absorb the movement (unless I was shaking with fear)? Unfortunately, I couldn't really determine which was best until I was in the air. I was a virgin (an aerial one, I told you who I married) and was volunteered because I had the Beta camera. I think the other reason was I would use less aviation fuel because of my smaller girth. It's nice to know you're good for something in life.

At 6:30 P.M. the next evening, our "ride" appeared. Landing in an abandoned airport three miles from our location was a 1949 Bell 47 D-1 helicopter. This patriotic, bubble-doomed bird of prey had been used as an aerial Jeep in the Korean War (seriously). The doors had been removed at my request to allow an unimpeded view of the earth below. The pilot, Jonathan Prox, had been flying since 1973 (boy, was he tired) and reminded me that a helicopter must be overhauled after every 1,200 hours in the air. Our unit was approaching its thirty-second time around. I asked him if he had a pop-up timer that was going to "ding" in the middle of our shoot. I didn't want to have to pull over in the air and change the oil.

Sensing my fear and apprehension as he tried to remove the camera from my whitened, clenched fists, he asked me what kind of shots I wanted. When I answered "ones in focus," he helped me into the cockpit. I was told to make myself comfortable and he would try to position the helicopter so I could get the shots I needed. Two extra tapes and four NP-1 batteries were wedged between my torso and my capable pilot. I could hang out, dangle, or peer out of the front or side of the aircraft as long as I didn't put my feet on the two pedals where I was currently resting them. It seems as if those nifty passenger footrests actually controlled the tail rotor and helped us steer. Wanting to be able to collect my pay for this excursion, I did as I was told.

I did feel extremely claustrophobic in the mobile fishbowl. I could see my wife Linda nervously snapping pictures of me as a tear rolled down her cheek. At least she would have a nice "last known photo" for my memorial service. Within seconds, we were aloft and headed toward our location.

I had to put my faith, support, and—oh yeah—my life in the hands of Jonathan. Instead of wearing the cool-looking headsets, I chose to scream my instructions to him as we flew. The headsets would just have gotten in the way as I tried to look in the viewfinder and I know I would forget to say "Roger" at the right time.

En route I tried to determine the best vantage point for the camera. Not wanting to miss a split second of in-air footage and the possibility of recording my own death, I maneuvered myself in the seat. My 34-inch waist had been reduced to a size 3 as I tried to slide in the seat without unbuckling the seatbelt, dropping the camera, or losing my upper dental plate in the gale-force winds.

I was able to get great shots with the camera on my lap, but my knees knocking added too much movement. As I hunched over the viewfinder like Quasimodo looking for a lost contact lens, droplets of sweat rolled off my nose onto the black and white screen. My new "for-distance-only" glasses were quickly becoming a pair of granny-style reading glasses as they slid down my perspiration-lubricated proboscis. Shoulder mounted was definitely the way to go. The human body does absorb most of the shake, even mine. Like Don Knotts at social functions, I was shaking like a Chihuahua on a December morning.

For a moment I wished I had a palm-mounted DV camera with its optical stabilizer, but no one wanted to throw one up to me. I guess I should use a better choice of words than throw up.

Once the camera was on my shoulder, the shot actually stabilized. I kept my nonviewfinder eye opened to judge the horizon and not to drench the pilot with the remains of my burrito lunch. As the 40-knot wind sent my tonsils farther down my throat, Jonathan mentioned that we were coming up on our first "swooping shot."

Since I was in my own little one-inch-square black and white world, I had no idea what Jonathan was doing to get the shot I requested. My vantage point was outside of the right "missing" door of the helicopter. If he flew straight, I'd have to shoot through the bubble cockpit and get instruments and a distorted view of the world. If I shot out the left, I would get the pilot in the shot. Jonathan knew what he had to do (with me still inside the helicopter).

From the ground, the producer, Tim Landis, and my soon-to-be widow, Linda, were watching our antics. As the chopping sounds of the blades were heard, the first glimpse of the helicopter appeared a few feet above the tops of the trees. I have since learned that the "chopping" sound the blades make is each individual blade breaking the speed of sound barrier, thus the "pop."

To help me out, it seemed that poor Captain Jonathan had to fly backwards and sideways so I would get the shot. He then sharply banked (without an ATM) and turned as I got a close-up of a woman (on the ground) expecting to be holding my camera in her teeth. The inexpensive approach to shooting in a helicopter is showcased in Figure 6-8.

When the viewfinder unfogged, I asked if we could repeat the maneuver, but this time remaining stationary over the audience and stage. Captain Courageous repeated the stunt and this time hovered over the audience. We now experienced two things I hadn't counted on happening:

■ If you are a major artist and are performing your act on stage, you usually expect it to proceed as it normally does without incident. It

Figure 6-8

Shooting in a helicopter without a Tyler mount

seemed no one told the entertainer we were shooting documentary footage from a helicopter. In the middle of his song, he looked up from the microphone and saw us hovering. Thinking we were the news media, he stopped and made some comment like "Where were you guys when we needed you in Vietnam?" and continued with his song.

■ It's very difficult holding a camera steady when a helicopter is hovering. The slower the air speed (we were stopped), the more unsteadiness in the helicopter. The wind was pushing us around as the pilot tried to keep the aircraft as motionless a possible.

In order to get a smoother shot, I asked him to fly over again and this time to go as fast as he could. *Never, never tell a pilot to do anything as fast as he can.* We started the next shot low over the river and followed its course banking to the left and right; immediately we were over the tops of the trees and heading for the main stage. The wind was deafening. I was sure that the last thing I would hear would be an angel singing. Colors were forming and blending on the horizon. The icy blast of the air froze the eyepiece to my cornea. (They say to get warm just before you freeze to death; I was now quite warm.)

As we approached Mach One, I started to black out. My backstage laminated pass around my neck was standing straight out and slicing at the pilot's throat. My hair had a permanent three-inch wide part, and the gray hair count had increased twelve-fold. The melodic angel's voice had now turned into demonic, maniacal laughter. Fearing that I was going to burnout in reentry, I held the camera as my knuckles began to bleed from the G forces.

After I glued my beard back on, I realized I had just shot the most exciting visual image of my life. I also learned that bugs actually scream right before they splatter on your lens.

Wiping the lens with my sweat-soaked shirt, I lifted the camera to my shoulder again. My muscles actually yelled at me under the strain. As we flew over the campground and parking lot, everyone waved at us. Why do people wave at you when you're in a helicopter? Are they glad to see you? Are they telling you it's okay to land by them? Are they trying to tell you your rear tire is flat? I wanted to wave back, but I would have needed to drop the camera first.

Forty more minutes in the air and we were ready to land. As they carted my pasty, bloodless body out of the cockpit, I looked like Don King before he combs his hair. At least I got the shot—alive.

The whole feat was quite a learning experience. You can't sneak up on anyone with a helicopter. Most of the shots I recorded were of terrified people. When something swoops down out of the sky, you naturally turn and look at it. You try being natural when "Kamikaze Chuck" is hovering there with a lens in your face!

The *Federal Aviation Administration* (FAA) also contacted me a week later because someone complained we were flying too low. Luckily, my wife had taken still shots of our antics in the air and after reviewing them the FAA realized we were flying at the legal height over people. Jonathan is also skilled enough to know the law and he wouldn't do anything to actually jeopardize anyone's life.

Other Avenues

As I mentioned earlier, I was able to find another market for the helicopter footage other than the documentary. Real estate people wanted aerial shots of the countryside for potential sites. The chamber of commerce also wanted my footage to use in a promotional piece about the area, and postproduction houses use some of the footage as stock aerial images (sell it to a stock library).

Never stop thinking about the potential markets for pieces or your entire project. As in my case, I did not own the footage I shot, so the producer had to negotiate any and all deals. If the footage is yours, leave no marketing spot untested. I never thought these markets would have any interest in my material; I won't make that same mistake again. Just make sure you know who owns the rights (if you don't know, ask).

The most important things to remember about shooting in a helicopter are these: Make sure you have a comfortable, steady shot; the faster in the air you move, the smoother the shot and the shorter your life; record everything—you can always edit out the shots of your feet, elbows, and

Review: Sixteen Things You'll Need when Shooting from a Helicopter

1. A comfortable camera. But for extended shooting in the air, a lighter camera is less fatiguing.
2. Get the widest lens you can find. Few zoom lenses are wide enough alone. Rent or borrow a wide-angle adapter. The increased field of view will steady your shots.

(continued)

(continued)

3. Have plenty of battery power. Extra batteries will burst into flames if they fall out of the cockpit.

4. Have plenty of spare videotape. Tape every second of your flight. You never know when you can use a takeoff or landing. At the steep price of renting a helicopter, make every second count.

5. Don't wear loose jewelry or sharp metal objects. You will learn a new meaning for the word "wind."

6. Color-balance the camera (if you can) in the air again. At various altitudes, the light has a different color balance.

7. Hold the camera on your shoulder instead of your lap. You have more mass to absorb any shakes. And don't take a milkshake up with you.

8. Tell the pilot exactly what kind of shots you're looking for. He or she won't tell you how to shoot them, but he or she will know how to make the helicopter accomplish what you want.

9. When the pilot banks right, lean to the right. Let the helicopter and your body be the tripod. Don't fight the feeling.

10. Keep both eyes open (unless your pilot closes his). If you close one eye and just focus on the viewfinder, you *will* get airsick. With both eyes open, you can still see the horizon, so your equilibrium won't complain.

11. Slower maneuvers will be shakier. Keep this in mind when flying.

12. Keep your feet off anything that looks like a pedal or control.

13. Make sure you're strapped in. If you lean too far, no one in space will hear you scream.

14. Get some shots through the front of the helicopter windshield; this adds a slightly different perspective to the shots.

15. Try not to zoom too much. Zooming is very unnatural when flying; you jar the viewer because the helicopter is flying at a speed unequal to your zoom.

16. Make sure you're actually recording the footage you want. Your mind may be on other things (like surviving) and you think you're recording when you're not.

Training Videos

Training the Untrainable

Some people just don't get the picture. I guess you might call them untrainable. But when you're asked to shoot a training video, it's your role to teach or train everyone. The slower learners may grasp the concepts at their own pace, but your video should educate all learning levels.

Tractors, Tractors Everywhere

I produced, shot, and edited a series of 38 training videos for New Holland America. Each video featured a different model tractor, baler, combine, or versatile. The finished program would be viewed by dealers across the country in the hopes of educating them on what a particular model did, its functions, and what to tell the potential buyers.

Most expect training videos to be dull, lifeless orbs with a narrator droning on about something no one cares about. If your subject matter is even slightly dull, you need to spice it up somehow. Some of the 38 training videos I did were not that stimulating in the area of subject matter, but we made them enjoyable by using a slight bit of humor, animated graphics (two-dimensional to keep costs down), a storyline that actually told a story, and kept the running time short.

An example from the unholy 38 was *Basic Hydraulics*. How do you make a 25-minute video on the concept of basic hydraulics interesting? Remember, you are talking about fluid, flow, pumps, schematics, and physics, so what should you do? You need to capture the viewers' attention immediately or you've lost them.

We opened with examples of hydraulics. By not just showing any example but an interesting cross-section, cut to the beat, the snoring would be less. I had to find a kneeling bus (one of those huge buses that hydraulically lowers on one side to allow people to exit), an open elevator where one could see the hydraulics in operation (a construction elevator), a car lift with an interesting antique auto on top, a car crushing machine that made a paper weight out of a 1979 Buick Regal, and a '63 Chevy Impala that bounced (low rider) on command (hydraulically).

Instead of showing dull things, we found hydraulic uses that were of interest because some were unusual. Showing the flow pattern of hydraulics is only accomplished through graphics and animation. By creating an intricate roadmap of open and closed valves graphically, something was still missing. Enter a comical superhero character, still animated, who would show how hydraulic fluid moves from one place to another.

Everyone likes comic books, so by making the superheroes humorous caricatures, we could have fun with the movement. For example, hydraulic fluid exposed to heat slows down. How can you make that more interesting and amusing? Show a large, green droplet of hydraulic fluid (looking like a Hershey's Kiss dressed in a red T-shirt) sipping a cool, umbrella-inserted drink through a straw at the beach lounging on a towel in the pleasant shade. Figure 7-1 displays the droplet in question.

We animated the fat droplet drinking by extending his lips an inch or two in one frame and cycling it back and forth. Sound effects also helped and his "slurping" sounds of an empty glass along with chattering seagulls completed the picture. We have just established how comfortable hydraulic fluid is when it's cool.

We then dissolved to a shot of the blaring sun and showed the hydraulic droplet as a shriveled-up, overbaked raisin a tenth of his original size with sweat pouring off his emaciated body. The point was made that hydraulic fluid does not functional well in heat. The droplet becomes rather thin in Figure 7-2.

Figure 7-1
A slurping hydraulic droplet at the beach

Figure 7-2
The dried-out droplet

Modulator Flow

In order to animate (with interest) fluid traveling through hydraulic lines, the superheroes entered the scene. A long line consisting of dozens of droplets marched into an open pipe. We animated the feet of the marching droplets and added troop footstep sounds. When a valve is opened, the marching pace quickens. This is shown by the superhero holding up a sign that reads "closed" and he immediately swivels the sign to read "open." The droplets speed through in a blinding blur and a ricochet effect accompanies it. By using animation to explain dull training material, the viewer becomes more involved, and that's what you want.

That Retiring Feeling

Another example of a training video was one I shot at a nursing home. I wasn't training people to be in nursing homes, but to educate potential inductees into what the place was like.

It's nice to shoot in a place were everyone calls you "young man." Since I don't usually hear that term much anymore, I try to savor it. Normally, when a client calls and wants a training video produced, I jump at the opportunity. For this client, I didn't jump quite as fast.

The client was the director of a retirement community, or in layman's terms, a nursing home. He said he wanted to dispel the myth that nursing homes were unhappy, dark places. He wanted to highlight the numerous achievements of his staff, how they knew all 600 residents by their first names, and how patient they were. An employee would spend an entire evening with one of the residents, often on his or her own time, if the resident wanted the company. I told the client that if a place like that really existed, I'd move in. After a long pause, I told him I was kidding and we'd be more than happy to produce a video for him.

I was working at a TV station at the time and this offered a few perks. Your vehicle has the station's call letters printed on every flat, semiflat, and glossy surface. That means if you park in front of where you'll be shooting, everyone within a three-mile radius knows you're there. Of course, we'd never use a vehicle like this to do anything illegal, like park 29 feet, 11 inches from a stop sign, exceed the posted 35 mph speed limit, or even think of doing anything that would discredit our employer.

When we arrived at the nursing, I mean retirement, village, we pulled directly in front of the main door. This isn't usually our practice, but the director of the village instructed us to do just that. We normally park the vehicle, go inside, and talk to our contact before we unload any equipment. As soon as I closed the door and turned around, three men in white coats blocked my entrance to the building. I smiled and said hello. The largest of the group pointed at our call letters and said, "Nobody asked you to come here!"

I rarely go to a client's office unannounced, and I began to wonder if I was at the right nursing home. Seeing the cameraman starting to unpack his gear sent the second goon over to him. "No cameras," he said. "Nobody here is going to be on the news tonight." The cameraman said that no one would be on the news tonight and continued to set up.

"I said we don't want any cameras in here," the gentleman said emphatically. I told him it would be very difficult to shoot a *training video* without the use of cameras. By this time, the director was on the scene calming down his agitated guards. It seems he neglected to tell them we were shooting that day. They weren't the first to notice our van and think they'd be on the eleven o'clock news. Our arrival had caused most of the residents to peer at us through their windows; so much for it being a friendly place.

The main purpose of the video was to tell the viewer how peaceful, friendly, and accommodating this place was. I certainly didn't have that impression from the guards. We had created a script, told through a voice-over, of how a daughter was taking her father to this retirement village. She had some early concerns, but once she saw the place, met with the management and staff, and saw the other residences, she was happy with her decision. This is very important in any training video: Tell the information in a story. If a person acts as if you're telling her story, the viewer will identify with the actor. Once that identification happens, you have them heart and soul.

Obviously, we couldn't use actual residents in the video because of releases and numerous other reasons. We hired two actors: a woman to play the daughter and an older man to play the resident.

Every new resident meets all the other residents, and the actors portraying the new resident and daughter would be no different. Here's when the fun began.

The first shot was to be a dolly shot from the perspective of the father entering the home for the first time. Since he was supposed to be wheelchair bound, we shot from that perspective. During the dolly shot, the director and the staff walk up to the new resident and introduce themselves. If that shot didn't look scary, I don't know what did. The camera is slowly dollying in at waist level, and here is this group of people in white uniforms walking toward you, grinning, with their arms outstretched. If I had seen that from the patient's level, I would have run from the chair screaming.

We reshot the opening from an over-the-shoulder approach as if we were one of the staff greeting the patient. This time the people walked toward him one at a time, instead of the Frankenstein mob approach. Through a series of close-ups, we identified all the key players and everyone looked happy.

The next shot was the more formal assessment each patient receives. All the key staff, the director, the new patient, and the daughter would meet in the patient's room to discuss his needs. The bedrooms were simple, tastefully furnished, but rather bland. I replaced all the lamp (practicals) bulbs with 200-watt photoflood bulbs. Of course, this made the lampshades glow like the surface of the sun. Carefully, I wrapped 216 diffusion over the bulb to soften some if its harshness. That worked until it would come popping out in the middle of a take. I was told every room only uses 25-watt bulbs to help the residents—help them go blind! The lamp shades didn't like the bulb's intensity either. We finished the scene quickly before the home was actually on the news that night—on fire!

The next scene was the new resident's physical therapy. The village requires that a family member be present at all therapy events. It's a familiar face to the patient and the family sees the care the resident actually receives. Our patient actor was so convincing, other residents kept asking him to be their partner in recreation, much better than an Academy Award.

The recreation area was the only place we were allowed to videotape actual residents. We could use no lights or close-ups, and couldn't stay in the room too long. We were not to call attention to ourselves in the large area. The facilities director figured that not having lights wouldn't call attention to us. He never thought about the huge Betacam camera we had, metal tripod, and our windbreaker jackets with the station's call letters stitched on the front and back like a billboard.

The staff set up the residents to play a volleyball game. They started them playing before we arrived in the room. We walked in quietly, set up, and I called action softly. As soon as our actor resident joined in the game, the other residents got mad. It seems some were offended that he turned them down as their partners. One of the other residents caught a glimpse of our camera and stopped playing. Soon others noticed him and they also stopped. It wasn't really funny, but here was the staff tossing this balloon (volleyball) and the residents would do nothing but stare at us. The balloon would bounce off them, but we were new, different, and much more interesting. Our actor, still trying to save the scene, wound up going after the balloon himself all the time. Needless to say, the scene was cut from the video.

The final scene was the tearful goodbye from the daughter. This also had to be carefully staged so it didn't look as if the father were being kidnapped by the staff. If she looked at him through the door, it looked as if he were a prisoner. We couldn't shoot everything from his point of view because he was in a wheelchair; all the staff looked too huge. We ended up just using close-ups of the father and daughter interlaced with more close-ups of them holding each other's hands.

It's really difficult to shoot a video like this with no budget and make it seem genuine. What looks great on paper doesn't always play out that way. My advice if you have this kind of shoot is to still plan it out on paper, but walk though everything before you have the camera. This way you'll get a better idea of what will and what won't work. And by all means, make sure the guards are told of your arrival. I still don't think they trust us.

Review: Ten Ways to Train the Not-So-Trainable

1. Add life to your training videos by showing examples of the training in action. You can't shoot a training video on assembling toothpick houses without showing toothpicks being used in that fashion. This may sound like common sense, but sometimes this simple fact gets lost in translation. Make sure people are seeing the training process in an understandable way. That's why they make training wheels for bikes.
2. Use graphics and animation to make the dull more interesting. Use characters from comic books or spoof famous people. If the viewers' attention is grabbed right away, it's easier to hold their interest than never having it in the first place. When I was trained for marriage, they attracted my attention by dangling a donut in front of me.
3. Don't have an on-camera person talking endlessly, explaining all you need to do. Instead of talking about it, show by example. This introduces the audience to the subject a lot easier (showing them) than having someone drone on about it. In school, those boring training programs only taught you how to make spit balls (not me, the other kids).
4. Become familiar with the subject you intend to train the viewer on. Research on the Internet, get facts from the company, and know the product as best you can. It's far easier to teach something you know something about. Maybe that's why men always develop commercials on feminine hygiene products with two women walking on the beach. Obviously, they know nothing about the subject. This may not be the best example, but I think you get the drift.

(continued)

(continued)

5. Determine early on what you want the viewer to come away with (don't train thieves; they come away with too much). Once the script is finished, have someone read it that knows nothing about it. Do they leave with the impression you intended? If not, use a shovel and make your own impression.

6. Tell the information (background, history, and so on) in a story. People seem to relate better when told anything in story form (look at *Toy Story*). If the trainee is entertained in the process, everybody wins. With a low budget, it's cheaper to entertain in the training video than hiring a juggling clown before every showing.

7. Use subjective shots wherever possible. These kinds of shots get the audience more involved and let them identify with what's going on. The only better way of training than subjective shots is to use a hands-on approach. Don't try this while on a date; here subjective is much better.

8. Put yourself in the viewer's shoes (once they stop wearing them). Being a novice, after watching the video, would you know what's happening? Don't talk down to the audience, but present the information in a way that explains, educates, as well as informs. If you can do that correctly, apply for a job where you shot the video because you have just been trained.

9. If shooting your video in an office, plant, or their location, blend in with the flow and don't distract the other workers. Anyone shooting immediately gets people's attention, but you need to focus on how to be invisible if possible. Wear a T-shirt that has no slogan on it and no one will notice you.

10. Plan everything on paper first before shooting any video. You may have to revise once shooting begins, but at least you have a game plan (unless it's Twister; then anything goes).

A Second Life for Your Videos

Videos, like cats, may have more than one life. You may possibly have created a video for a client, it has done its job, and now someone else may want to use it. I shot interactive dental training videotapes at a community college centuries ago. An Apple II computer accessed the VHS tape and took the dental student through tooth scaling, pit and fissure sealants, and determining and helping prevent gingivitis (I sound like a Listerine ad).

These same interactive videotapes were transferred to DVD recently and students many years later are watching the same training material in a new interactive form. I did the same thing with a criminal justice video on police training. Beside the out-of-date hairstyles, the techniques haven't changed, only the presentation medium. These videos have lived again (I should begin maniacal laughter).

The filmmaker Sam Peckinpah once said, "A film is never completed; it is only abandoned." He's quite right with that statement because a video never really dies or is useless; it just may take on another form. Like the dental video, new life may be breathed into an old video, making you profit. How do you make a video with the future in mind? A lot of videos shot in the '80s look bad because of the lesser technology. When I shot back then, I used the best production values I could. Make sure your production values are high now, so the video will stand the test of time. One way of making that happen is with filters.

Working Without a Net

Video has a beauty all its own, but the recorded images can be enhanced. It seems when people are shooting video, they want to make it look more like film, and film shooters want to harden the image to make it look more like video. We never seem to be happy in the medium we are using.

Video does have a realistic, hard edge that some will try to soften. We recently had a client that wanted the final video image to look as far removed from video as possible (so it could be used far into the twenty-first century). The easiest way to accomplish this was using glass and a net. When I was in the circus, I never worked with a net, but this is a good place to use one.

The first critical step in shooting anything is with your lighting. We were to have a woman being interviewed on camera. The client wanted a soft look with soft lighting. Our key was a Chimera with two 1,000-watt Totas, each gelled with 1/4 Booster Blue and Tough Frost. Through the Chimera's white fabric surface, the key light was very diffused. The fill and back lights were Arri 650-watt Fresnels silked with Opal Diffusion and 1/4 Booster Blue.

If you want to further soften an image, you should start with soft lighting. I found that the Chimera is an excellent source of soft, directional lighting.

To further soften the look of the video image (and make it look more like film), we shot through a black net (stockings). A black net stretched in front of or behind the lens will break down the harsh, sharp video edges and blur them slightly. This is the softening effect the client wanted. It's almost as if every pixel in the chips now had rounded edges instead of square.

Black hose was chosen because it acts more like a neutral-density filter and doesn't affect the color of the shot. If flesh-colored stockings were used, the flesh tones would have a much redder or warm appearance. White stockings would have paled the flesh tones. Any other color stocking would add that color to the shot. Black stockings leave the image as it actually is, only softening it greatly.

The type of stocking chosen is also important. If I went to the local discount store and bought a cheap pair of pantyhose, I'd be in trouble (from my wife too). When inexpensive pantyhose is stretched to be put in front of the lens, a few things usually happen. Because the product is cheaply made, it will tear or run easily. You don't want to shoot through a rip, tear, or run. The fabric used is also not as seamless. You want to shoot through a sheer section, not one full of bits of fabric clumps.

Our *director of photography* (DP) always buys his stockings (for the camera) from Frederick's of Hollywood. They are more expensive than conventional stockings, but are well made, very sheer, and any section of the hose can be used over the lens.

Is it better to place the pantyhose over the front or the rear of the lens? That's really your call. If you put it over the rear of the lens, you will have a much smaller working section of the fabric to shoot through. In addition, the remounting of the lens can tear the fabric, destroying the effect you were after. I prefer stretching it over the front of the lens. This gives me a larger working area that is in less danger of being torn. People usually question me when they see a leg of black pantyhose dangling in front of my lens, but after they see the image it creates, they will soon want one dangling over in everything they shoot.

I walked into a Victoria's Secret (we don't have a Frederick's) and asked to buy one black stocking. The cheerful woman said, "One pair in what size?" I said I didn't need a pair, only one leg. That's when she rang the buzzer for the manager. One leg stretched over a lens is enough netting to last eight years. When a section wears, move over a few inches and use a new piece. I ended up buying black stockings that stopped at the knee (enough diffusion for my needs, and I still had to get two).

This pantyhose trick is nothing new; photographers have been doing this for years. You also don't need any special kind of camera; stockings can be used with video or film cameras.

Because you are stretching something black in the front or rear of the lens, you are going to lose light. With the Frederick's stocking, you lose about a stop and a half, but the overall effect is amazing. The image is still clearly focused, but all of video's hard edges are softened and blended. The on-camera image tends to glow slightly.

Two other ways of achieving this same type of look are with a Pro Mist filter or petroleum jelly. Tiffen makes a great Pro Mist filter that softens the image slightly (less than the hose and allowing more light to enter) and doesn't look as strange as having a leg of pantyhose dangling from your lens. Every time I shoot with digital video, a Pro Mist filter is always involved.

Gooey

Vaseline (petroleum jelly) can also be smeared on a piece of glass and placed in front of the lens. Never smear anything on a lens except lens-cleaning fluid; it's difficult to remove. Instead use a piece of glass and carefully put a light coat of the jelly on the surface. It's best to leave the center area clear and the edges blended. Put this "Vaseline-smeared" glass in front of the lens and shoot through it. The use of petroleum jelly takes a lighter touch than I have. I usually put too much on at a time.

My, What a Long Lens You Have

Now that the image has been softened and has a romantic, heavenly glow, it's time to move to the next step in the softening process: the long lens. We shot our subject with a Fujinon 8-110mm lens. The camera was placed at the rear of the room and zoomed to its most telephoto position. This also helps blur the background behind the subject. Telephoto lens compress or soften the image. The client wanted the subject to look as if she were in front of a blurred background, much like a lone flower in a field of grass. To further enhance the softness and blur of the image, the Fujinon lens was placed in doubler mode. This made our lens a 220mm unit and the doubler softened the image even further.

The Fun House

The third type of effect to create our desired image was to place pieces of curved, wavy, or beveled glass in front of the lens. This step also has to be done carefully. If you place a piece of wavy glass directly in front of the lens, the entire image will look strange and distorted. Rather just use an edge of the glass in the frame. This edge will distort or change that portion of the image. The subject can be in clear focus in the center, but various pieces of glass on the sides, top, or bottom can add to the effect. Figure 7-3 shows an example of glass positioned in front of the lens.

Figure 7-3
Wavy pieces of glass

Figure 7-3
Wavy pieces of glass

These pieces of glass can be attached to arms on C-stands and placed at the desired angle. Try shooting using an edge of the glass in the frame, or even let one of your lights reflect off one of the pieces to add highlights. There is no right or wrong way to do this. Experiment and see what works best for you.

Now that you have the image exactly the way you want, it's time to shoot it. Our client wanted a slowly moving camera combined with a Dutch tilt. As the on-camera talent answered the offscreen questions, the camera slowly moved to the left and then right. The movement should be extremely slow, not to call attention to itself. Other schools enjoy using the snap zoom method combined with the oblique angles. Do whatever you want and experiment. Once again, there is no right or wrong way to do this.

Arrive at your next shoot with pantyhose and several samples of glass. Try them out for the client and see if it's something they like (you know what I mean). These effects aren't for everyone, but they can enhance the look of your next training video and may give it another life.

Bits and Pieces

Don't be afraid to use parts of your previous videos in new ones (if you own the rights). You can be your own stock library. When I need aerial footage, most clients don't want to pay the expense of sending me up in a plane or helicopter. If the footage can be nondescript, I'll use some of my older aerial footage.

Shoot every video like it's going to be archived or reused at some later date. This is the epitome of no-budget production. Your only expense on the new project may be editing the old footage into your new video. This means you should keep, not erase, everything you do. Boring shots of your left toe when you inadvertently left the camera on while walking are useless, but any well-framed or planned shot could be used again.

The biggest mistake in low-budget productions is the belief that recycled tape will save money. Don't get me wrong; it is cheaper to use someone else's tape stock, but once you record an image, save it. The tape's recycling days are over. In a short period of time, you will amass a library of, well, *catalogued* footage. This does take a few extra minutes in the early phase but will save you tons of money later because you have the shots and won't have to reshoot. I'm always getting calls from producers who ask if I have any shark attack, police chases, or strange events recorded on tape they can purchase for the reality shows on TV. Unfortunately, most of my footage (video footage) is too mundane for them, but it was nice of them to ask.

A colleague of mine was bidding on a video shoot in high definition. Her selling point to the client was because we were still shooting on tape rather than film, that the high-definition tape could be erased and used again where film, once processed, may not. After I got finished choking her, I explained that having anything on high definition today can be used for eons. High definition (at the time of this writing) is a hot format and people are clamoring for the images recorded onto it. Don't shoot yourself in the foot and erase it.

The only way to give your videos a second life is to save everything; catalogue and label it everywhere, always. Search for new avenues for its use, keep your production values high without the expense (use good lighting, filters, and so on), and when new technology surfaces (and it will), see if any of your old programs may fit the bill.

Review: Seven Ways to Make Your Training Video Unique

1. Give your video a different (better) look. This may be achieved with lighting effects (soft or hard) and filters. This will not work if you hold an oil filter up to the camera.

(continued)

(continued)

2. Use a stocking to blur your image slightly. This adds a romantic look to your video and people might not expect this type of "class" in a training program. Just remember to explain why you have something dangling from your lens (make it interesting, okay?).

3. Another softening approach is to use a diffusion filter (Pro Mist) or petroleum jelly on glass (pheasant under glass tastes better) in front of your lens. This creates a more diverse look than stockings, which is why people don't smear petroleum jelly on their legs.

4. To soften the background, use a longer lens, a doubler, or open your iris. This makes the subject stand out more. The only other approach is to get a blurry background and don't wear your contacts.

5. Stick a piece of glass into the frame slightly to add interesting visual elements to the shot. Unless your training video takes place in the hallucinogenic 1960s, don't shoot through the glass.

6. Angle the camera (not as far as a Dutch tilt) to change the perspective slightly. Once again, *Batman* used this technique to his advantage on TV, but where are you going to find "Whap," "Zowie," and "Crunch" graphics?

7. Use the footage you shoot to compile your own stock image library. By carefully cataloguing all your material, you'll have less expensive images to use where you like. Just leave little princess' second-grade recital footage out of the library.

Don't Forget the Government

I've received more training work from the government (ours, that is) than any other source. I've traveled this great country, shot in exciting ports of call (Army bases), and worked with some pretty sophisticated equipment (all obsolete now). These training videos, although done with primitive technology by today's standards, are still in use training our armed forces.

Your Pad or Mine?

Back in the mid-80s, I produced 12 interactive videodisc programs for the U.S. Army called *The Positioning and Azimuth Determining Systems* (PADS). This video program instructed personnel where to position their gun batteries while using a positioning display system in their Jeeps. Much like a global positioning system of today, the soldiers could drive from point to point and the PADS unit would tell them went to stop.

All the video material resided on a double-sided videodisc, 30 minutes per side. A computer with two 3¼-inch floppies would branch the user to different parts of the program while keeping score when the touch-sensitive TV monitor was activated. The same program today could fit on two dual-layer DVDs without the need for a computer.

The government still produces numerous video-training programs and needs talented video personal. All branches of our Armed Forces have in-house video people, but the sheer mass of projects necessitates them to hire outside contractors. My first job with the government was shooting one-inch tape and editing on two-inch quad, which soon changed to Betacam SP, and now I'm using digital for all my government work.

Luckily, the government changes with the times almost as fast as the industry. They need to use cutting-edge technology in an effective way to train our troops. When doing work on aircraft carriers in the mid-nineties, video equipment would not function because of the magnetic interference from the radar equipment. All training had to be done on 16mm. This may have changed.

When looking for government work, the *Commerce Business Daily* is a great resource. Also available on the Internet, this periodical lists every contract the government is bidding, video and otherwise. The key word here is "bidding." This is an area where the lowest price wins (with professional capability). Don't expect to land a government contract when you bought your first camcorder yesterday and aren't sure which end to look into. The cheapest price will win, but you must have the skills necessary.

Some contracts require you to be on the Qualified Video Producers List. Once you are on this list, your video expertise is deemed acceptable and you are allowed to bid on any video project. Getting on the list only involves sending a tape of your best completed work to the government where they will review it and give you a grade (just like school).

What better place for no-budget videos to become a reality? The government has tons of contacts, but not billions to spend on them. In the '80s, we had million-dollar contacts; that era has ended. With incredible competition out there, I've seen winning bids of $12,000 on 30-minute videos. These projects may never finish, but low-balling a bid to get in the door is not the way to win new friends.

On current government contracts I have bid on, they ask for an hourly, daily, and weekly rate for each crew member (director, camera, and so on). If they have a 30-minute video open for bidding, they just want you to fill in the salaries; they will determine how much time it should take to shoot (they have the experience; they should know). Instead of saying you can unrealistically shoot a 30-minute video in two days and give them a crew price of $12,000 (a great price at that), they will do the math, and with a crew and equipment bid at $3,000 a day, seven shooting days for $21,000 is more realistic. This new structure covers everyone much better.

If the Water's Green, Don't Go In

Sometimes you're invited to shoot things that you would do even if it cost you money. I was recently able to take part in the *Comprehensive HAZMAT Emergency Response Capability Assessment Program* (CHER-CAP) training exercise in sunny Chesapeake, Virginia. I had more fun on this shoot than I've had in a long time.

Being the largest exercise of its type in the United States, I was one of the five video crews that captured his historic training event on tape for the *Federal Emergency Management Agency* (FEMA). After 8 months of planning, the CHER-CAP exercise used 500 Navy volunteers as "victims" of 2 separate terrorist attacks. The actual program allowed firefighters, medical, and *Hazardous Materials* (HAZMAT) crews the opportunity to train in a real situation with the actual horrors associated with it.

I showed up bright and early on a muggy summer morning on August 18, 2001 (the date is important). With the 9:00 A.M. temperature already a heady 94 degrees, even the flies were too hot. Immediately, I was handed a red vest that said "Media" and I would be allowed to wander anywhere at will without being shot. I really wanted a vest that said "High Exalted Grand Poobah," but no one offered. As long as no one asked me to stop traffic and set up plastic cones, I was happy.

The Army and Navy also had handed out red vests for their media crew (we could also choose up sides if anyone wanted to play an Army/Navy game, but we wouldn't know which team was which). Several others were issued the same red media vest but with the work "Invisible" written on it. First of all, writing the word Invisible on a piece of fabric doesn't allow you to hide in the girls locker room (no, I didn't try it). Invisible means that these individuals would be recording the event for a separate videotape (other than some news media folk). The actual media people had the word "Real" on their vests (I was just a fake person). You just weren't supposed to pay attention to anyone with Invisible on their vest (like how I felt next to all the tall girls in the locker room . . . oops!).

I had finally reached the big time and would be shooting with the talented folks of the U.S. Armed Forces. I, however, was given a slightly different role. I would be capturing the day's events on a prosumer Hi8, a Sony TR615. Before you email me and call me a neophyte, there was a very good reason for this. When an actual catastrophic event happens (like September 11th, then less than a month away), news crews cannot be immediately on the scene. Sometimes a lone, bored pedestrian happens to live nearby and captures the event on his or her home camcorder. That was my role. This home camcorder footage would be intercut with the Betacam footage to show the grittiness of the destruction compared to the reporters giving their side of the story.

I happened to be first on the scene before the news crews arrived. I had to record the screams of agony and the first initial cleanup and detoxification events on a prosumer, one-chip camera. This footage would add realism when eventually edited with the Beta footage. I was still a professional with a prosumer camera, so all the proper angles, cutaways, and sounds would be recorded, just on a slightly lesser format. It's really not the tool you use; it's the knowledge behind it. If you don't believe me, you should see the book-case I built without using a saw.

The first event involved a 30-car train that had stopped in the middle of a city park. Over 150 people were in the park when a small charge punctured a hole in a tanker car. After I got off the ground from the concussion of the boom, I walked over with my nifty vest and began videotaping.

Human body parts (mannequin appendages) were strewn all over the train tracks and 20 people were lying in the grass with burned faces and mangled appendages, yelling for help. The actual tanker car contained 50,000 gallons of cooking oil, but black gaffer tape had been used to hide this actuality (I could insert a joke here about cooking oil, but I'll let it slide). A garden hose was taped to the metallic side of the rail car and spewed green-colored water, showering the victims. The simulated leak was supposed to be hydrochloric acid spilling. Figures 7-4, 7-5, 7-6, and 7-7 show images of the CHER-CAP exercise.

The large green men arrived and began plugging up the leak. I got in people's faces with my camcorder, showing the agony of what had actually happened as they asked me for help. As a news gatherer, do you stop and help them or continue to capture the event (that no one else had recorded)? It's a difficult decision to make, and I would be torn if these people had actually been burned with HCL.

A mob of Army officers in gray spacesuits arrived to help the victims. As I calmly wandered among the human carnage, I got fantastic footage without being too artsy. My Hi8 camera was much lighter to hold, I got two hours on a tape load, and if I had to leave the scene because I was glowing (no relation), I could always jettison the camera after I extracted the tape.

Figure 7-4
Shooting CHER-CAP

Figure 7-5
Spill cleanup

Figure 7-6
Behind the scenes

Figure 7-7
Water spray detox

When the Beta crews arrived with their 26-pound behemoths, I had already recorded the immediate aftermath of the incident. The fire department turned on their hose to detox the victims still on the scene. Water is the most common way to remove acid from the body. As the Beta crews tried to cover their lenses from the onslaught of high-pressure water, I calmly shielded my tiny lens with my hand. I know knew why I had been chosen to record with the one-chip model; I could run faster than the guys with the heavier cameras. A lighter, digital, three-chip camera would only have added a few pounds and greatly increased the quality, but grit was in for this shoot and it kept our shooting costs down.

The victims that could walk went towards two fire engines that had a tower of water extending high above. This "shower" would remove any possible chemicals and detox the individual. In a brief moment of madness, I removed the large baggie from my pocket (my lunch had been in it, okay?) and covered the entire camera. I walked behind the hordes entering the rainfall and stopped just at the edge. I got the only subjective shot of the water tower. By the time the new crews had arrived, I had already dried off.

The second scenario involved a Dodge minivan being blown up by the same group of terrorists. I have never seen a van so mangled from something as small as a pipe bomb. I walked along with the firefighters as they helped carry the survivors to safety. With Sony's Steadishot, I could walk at a fast pace without the usual camera bobbles that are associated with this type of movement.

Just like an actual event, the place was teaming with reporters and camera crews. The authorities on the scene had to keep telling the media to back off because they were getting in the way of the triage procedures. In order to get the shot, you need to stick your camera where it sometimes shouldn't be (watch it). However, with my prosumer model, I could wander right into the heat of the action and people thought I was just recording with my home camcorder (except for the red vest). A smaller camera is more invisible at times like these, and no one asked me to leave, something that had been done with every Beta crew.

The moral of the story is that I actually did a better job covering the event with a prosumer camera. A large unit would have been more fatiguing and not enabled me to get as close to the action as the director demanded. It's always best to do what the director wants; his or her tantrums usually last longer.

On September 11th, every member of the HAZMAT team was called into action for the terrorist attack on the Pentagon. This training was all too timely. The mistakes learned in August helped save valuable time in September.

The controlled exercise on tape is obviously a much safer way to deal with the unfathomable number of casualties that September 11th brought about. Personally, it means more to me in that I may have contributed in some small way to help these gallant men and women who are risking their lives for our safety.

Review: Ten Reasons Why Smaller Is Better at a Live News-Type Training Event

1. With a small camcorder, you can easily blend into the action and no one knows you are really video savvy. A prosumer camera sometimes may mean someone not as experienced is behind the viewfinder. In order to avoid confusion, I wrote the word "Professional" across my vest (but I spelled it wrong).

2. When capturing gut-wrenching events, don't look into the viewfinder. If your camera is set on a wide shot, point it in the general direction and you will capture the shot. This way people don't think you're videotaping them and they won't be staring at the camera. This doesn't work when you shove the camera two inches from someone's nostrils; they will look into the camera just before they hit you.

3. Cover any flashing or steady, red record lights on your camera with tape. If people don't know it's on, they may be less apt to give you the "deer in the headlights" look. One professional not having tape to cover his light used a Mickey Mouse Band-Aid instead; that actually got more attention (probably because it was an extra bandage, not because it displayed a picture of a four-foot-tall mouse).

4. If you need a subjective or *point of view* (POV) shot, make sure the camera has protection (not muscle-bound men with bent noses). Water, dirt, or sand in the wrong places on a camera is trouble, and tough to explain to your spouse.

5. The fatigue factor is much lower carrying a microscopic camera. When sweating is the main activity of the day, it helps when you have a smaller camera. Last time I had a Betacam, I knocked myself out when I tried to wipe the sweat from my brow (forgetting I was still carrying a 26-pound camera).

6. Victims generally feel more comfortable with a smaller camera around. A Betacam means "the news" and some people don't want to be seen that way on TV. Most will allow you to get closer with a camcorder because they know few will be seeing the results (your mom and grandmother). Slipping them a 20 doesn't hurt either.

7. In the same light, I captured unique behind-the-scenes footage because I had a camcorder. I actually videotaped the FBI getting dressed for this event (they don't have the letters FBI tattooed on their skin like I had heard). When a Beta crew arrived, they were asked to leave because the FBI said, "No cameras, we don't want this videotaped." Of course, I actually recorded him saying that on my camera, so if that tape ever surfaces on the Internet, I'm a dead man.

(continued)

(continued)

8. Twice I noticed the Beta crews having to stop and change tapes in the middle of an interview. With two hours of tape, you won't get a break in the action. The downside is people ask you to play back what you just shot. They never ask that with Beta people. Maybe it's the cap I wear with the word "Tourist" on it.

9. Don't be ashamed if you are asked to shoot with a lesser camera on a project. You can now really show your talent because your skill will show up better than most images. Just make sure you take the lens cap off first; that goes a long way.

10. Don't make friends with the victims in a mock HAZMAT exercise and tell them you work for *Life*. I actually meant I worked for the people who made the breakfast cereal.

When Do I Join Forces with Others?

It's not easy to do a no-budget production all by yourself. You only have so many hands and you may need assistance with your project. This is nothing to be ashamed of, but it may add a few dollars to your budget. If you are incredibly stingy and do want help without paying for it, you only have two avenues open: Get students (interns) or trade your services.

Summer Help, Some Aren't

A world of free labor exists out there called students. They can range in age from high school to senior citizens leaving the work force. Just because your help may be an aspiring video maker in the guise of a student, don't take advance of his or her naiveté. As I mentioned earlier, my students work harder and are more professional than most people in the business. This eagerness to learn makes them great assets to any production. Most times the experience gained on a shoot is more important than money.

Interns on your production may not be available if you aren't working for a company or will not give them the "credit" they need for their particular course. At the TV station, we always had a flock of students who acted as interns during their junior or senior year of college. Most interns work for college credit and aren't paid. Unfortunately, they work harder and longer hours than most employees because "they are learning" (this is when some of them learn not to go into the video business).

My only responsibility as an advisor or supervisor to the interns was to make sure they had work to do, and I had to evaluate them at the end of the term. This is great place to start out in the business and, if already established, a great place to get labor. Talk with local colleges or high schools and see if they offer credit to students who want to intern.

Don't Get Sick on Me

Since you're totally in control when shooting a video, why not make the viewer experience the same feeling you have when you sit up too fast? Have you ever seen to shoot something with the camera pulling back and simultaneously zooming in? This "move" hits you in the pit of your stomach because the object isn't moving, but the camera is zooming in while dollying back. The background seems to move, but the subject stays stationary. This effect was done in the movie *Jaws*. Roy Schieder is sitting in the boat when he sees "Jaws" for the first time. The camera dollies back and zooms in at the same time—giving the viewer the sickening feeling Spielberg intended.

Wanting to achieve this same uneasy feeling that so many filmmakers have successfully accomplished in the past, I set out to shoot it (only I had no budget). Making the complicated shot even more difficult was that the action would take place in the backseat of a car. I tried getting my motion control rig wet so it might shrink and fit inside the car, but all it did was rust. We had to resort to "the cheaper, makeshift route" to shoot the shot.

The actor, Billy, was to be sitting in the backseat between his girlfriend and her eight-year-old sister when he realizes that the entire family is nuts and he wants out. While holding a milkshake in his hands, he begins to quiver, tremor, and shake as his nerves get the best of him. The camera then zooms in and dollies out at the same time to make you feel what Billy is going through.

Because the rest of the video had been completed, Billy was brought back to a school parking lot so the zoom-dolly combo could be created. Racking my brain to come up with an inexpensive method to complete the shot, I thought it might be a good idea to invite my video production students along to watch me in the midst of my madness.

Twelve eager students arrived on a foggy afternoon to recreate history. Although video reacts poorly to fog (it looks like out-of-focus video), because the action took place in the car's interior, the diffused background posed no problem.

The milkshake was assembled using a mixture of shaving cream and water. This soapy-smelling concoction was for quivering, not for consumption. The camera would be attached to a piece of plywood and this dolly could then slide along an aluminum channel with wheeled castors (or in layman's terms, on a skateboard sliding down a rain gutter). The front edge of the smooth aluminum channel rested on the windshield wipers and the back end was elevated nine inches.

When you're shooting on the hood of a car, you look for anything that will give you elevation on the slope of the internal combustion auto. An empty camera bag was placed under the front end of the ramp and a metal camera case, blankets, and a Betacam cassette case rounded out the rear (I just watched a *Buns of Steel* video). Because we were shooting down into the car, the dolly platform must slope toward the windshield.

You only need a few other necessities to finish your dolly track of death. Cover the car's hood with a blanket so the finish won't be scratched and your butt doesn't freeze to the surface. With a semicomplex shoot like this, you should have some sort of monitoring system to view the dolly/zoom. Without the aid of a flip-out screen (I'm glad nothing else flipped out on the shoot), RCA cables were attached to another camcorder's monitor. The director and students could now watch me fumble my way through the shot.

Since most windshields are curved slightly, a Tiffen polarizing filter was connected to the lens and rotated to eliminate the reflections on the glass. The rear window behind the talent was gelled with Rosco ND9, and silver reflectors were placed on the left and right of our actor. The rear side windows were lowered to allow the fog and light into the car (side windows are also tinted), and the front of the car was covered with a blanket to prevent the drizzle from hitting the glass. Four tall students each held the blanket on a corner to keep my hair from getting frizzy. Images from the hood shoot can be seen in Figures 7-8, 7-9, and 7-10.

Climbing on board the hood, I placed myself near the lens so I could practice my zooming in as the camera pulled back. Originally, I intended to dolly back and zoom in myself, but my coordination isn't what it used to be. One of my students would dolly back at a predetermined speed on my command. As I counted from three to one, she would dolly back, hitting her end mark each time. I was so impressed at her preciseness that I neglected to zoom in.

The key to such a shot is the timing. If the dolly is too long or short, stops too soon or late, or the zoom isn't the precise speed, stopping and starting at the same millimeter, the shot doesn't look right. Although mildly difficult, good timing can be achieved if you practice and have your moves synchronized. They say practice makes perfect and that's exactly what you must do.

Figure 7-8
Camera setup

Figure 7-9
The apparatus

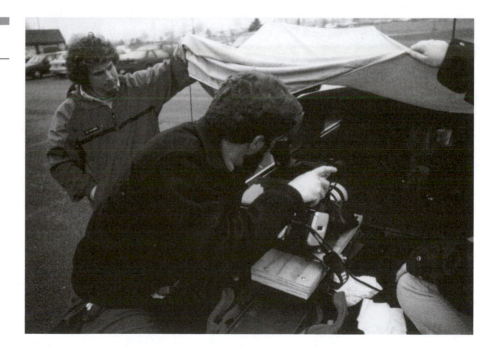

Figure 7-10
Camera on a
skateboard

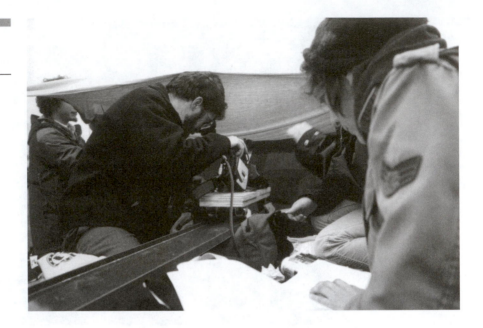

This was an inexpensive shoot that was accomplished with the help of students. When you are trying to save money on a shoot, you always seem to need five times as many people. It's a good thing they were students.

Swap Meet

The other option of getting free help is to trade services. You work for someone on their shoot for nothing and they return the favor. Sometimes you can't put a price on someone's expertise.

Federal Express wanted a training video on the process of de-icing an aircraft. I had to travel the country at a moment's notice, watching for snowstorms. If one was scheduled, I needed to determine where Federal Express had their planes, then jump on one (a commercial airline), and get there before the storm it (luckily I was shooting this in the winter).

In the warm confines of my home, Burlington, Vermont (two hours away by air) had predicted a doozy of a storm that afternoon. I love Vermont, but you can't get a direct flight to Burlington from anywhere except Burlington. I packed my stuff, rented a Betacam from a local rental house, and boarded the plane.

When I arrived in Newark, New Jersey, I had to change planes to get to Burlington, the storm being two hours from happening. This was a few years ago when Peoples Airlines was still in business. This was the airline where you actually paid for your ticket while seated in the plane. My expensive, rented Betacam had to sit in the cargo hold because no overhead bins were available. After waving a tearful goodbye, the camera case with the words Betacam disappeared behind the conveyor belt's fabric flaps into oblivion.

When I arrived in Burlington, the snow had already begun. I just needed to grab my camera, get a rental car, and dash to the Federal Express hangar. I waited over an hour, watching the blizzard coat the roads, and watching for my camera. It never arrived. Later I learned that someone seeing the words Betacam thought it was a new camcorder and probably stole it; it never made it to Burlington. I'm sure they had no clue its value was $30,000.

That brings up another point. No matter how low your budget, *always* get insurance on rental equipment. My boss, who owned the company, traveled with me and would not pay the rental company for an insurance rider. He now owed the rental house 30 grand, a very expensive lesson learned.

But my biggest problem was being in Vermont, in a snowstorm, with a shoot happening in less than one hour, and no camera. I called every video house in the yellow pages (three of them at the time) and pleaded my case. One guy was available who said I could use his camera.

Talk about how great and trusting people from Vermont are. He drove to the airport in the storm with his only camera and let me use it—at no cost! He later told me that when he was driving up to the terminal he saw my tear-stained face, quivering lower lip, and snowflakes sticking to my beard, and felt sorry for me. He said I looked like I could be trusted. If I had not passed the "he looks okay" test, he told me he would have kept on driving (since we had never met, I would have had no clue that he had passed me by and I might still be standing there).

The snowstorm shoot went well. I huddled in a cherry picker bucket 30 feet above the runway as the de-icer hosed off the tail section. I had frozen my tail section off so this was the next best thing.

This is another great lesson about the video community. This guy did not have to do what he did for a total stranger. I offered my services to him if he was ever in my area. I'm not saying you should "give" your stuff to a stranger, but offering your services in exchange for theirs will open new doors. They may get you on a big shoot that you would never have had access to without them.

Whenever I travel and need a video crew, I use the Internet to find capable people. When we meet in that strange town, I have local help that can grease palms for me, I don't get lost driving to different locations, and I have new friends that share my interests. Depending on the budget, each and every professional has adjusted their rates accordingly for the chance to work and the experience.

Enlisting help from others might also help when no one on the shoot knows how to get a particular shot covered. The local guys have often saved the day by making a suggestion that we ended up using. If the roles are reversed, offer a suggestion if you think it may help.

One last thing I can say about training videos: They have all educated me. I learn more on the set of a training video than I do on a corporate video, documentary, or commercial. I believe it's a learning process because you are teaching someone how to do something (simply) using video, and a benefit from this process is that you've added another trick or skill to your repertoire. Just remember, when you stop learning, you're dead!

Review: Six Tips for Traveling with Equipment

1. Arrive at the airport or final location well before you are expected to shoot. Too many things can occur when you are responsible for equipment and yourself. Look how much trouble it took to find that special toothbrush.
2. Carry the camera with you when traveling. The case may sit with the rest of the cargo, but something as delicate as the camera should be in your possession at all times. This way if there's an accident, they can identify you because you will have the camera clasped to your hands.
3. Get insurance on all rental equipment. Most times it's mandatory, but the small expense is worth the protection. I said the same thing about "Luigi the Knot" earlier.
4. Wherever you're shooting, have a listing of backup resources. Murphy's Law only happens when you are least prepared. Even if you don't end up using the personnel, they can help steer you away from the seedy places (where they make rye bread) after the shoot.
5. Carry spare everything with you. This adds a little more weight to your bag of goodies, but if it's there, you'll use it. This does not apply to clones.
6. Hire local people (local, not loco) when traveling out of town. They know the town, the people, and locations better than you do. They may return the favor when they come to your town (so don't stiff them for the parking meter money).

CHAPTER 8

Interactive Programming: DVDs

DVDs: The Sordid History

Hopefully, you already know what a *digital versatile disk* (DVD) is. We rent them from a local video store and are thrilled because we aren't charged a dollar because we forgot to rewind them. DVDs have become a boon to the no-budget video producer. Your best video work can be recorded on DVD and shown as part of your demo reel, used as a promotional piece, and sold internationally.

DVDs were the brainchild of the Hollywood film industry as a way of getting new releases (of high quality) out into the hands of the public. When the Academy of Motion Picture Sciences is evaluating new films for the Academy Awards, all the entries are given out to the judges on DVDs. This DVD is an advance copy that is pressed sometimes only days after the theatrical release. This is far easier that inviting everyone to a screening and hoping all can attend. By sending the discs out in the mail, each judge may view the film privately at his or her leisure.

I'll talk about two different kinds of DVDs in this chapter. The recordable DVD formats will be discussed in this section, and the mass-produced version will be in the last section ("Reaching Millions by Spending Thousands").

Smile, You're Being Recorded

With the advent of *DVD recorders* (DVD-R or dual-sided DVD-RAM), almost any program may be recorded or burned onto a blank DVD. A DVD-5 holds approximately 5 gigabyte of material (4.7 gigs) and a DVD-9 holds 9 gigs. The DVD-9 employs dual-layer technology, which encodes two different layers of material on one side of the DVD rather than flipping sides like a Herman's Hermits record. You've all seen this layer change occurring: In the middle of an intense scene, the image freezes momentarily as the laser tracks the second layer. In some films, this layer change is more cleverly disguised and unnoticeable in a fadeout.

The average DVD burner can record information from any tape format to DVD. A favorite TV program can be time-shifted and recorded on DVD like you would on VHS. But you're probably more interested in how to get your valued production onto DVD. It's not rocket science to simply record from tape to DVD; anyone can and has done it. But what if you want your DVD to be interactive, like the Hollywood features?

That's where menus, branching, and interactivity are all lumped into the word of the day, *authoring*. Authoring enables you to create a simple or elaborate method of making the DVD do exactly what you want it to do. The viewer can now watch any portion of the program in any order. In training, the result of a choice the viewer makes may take them along a different path to an alternate conclusion.

In my criminal justice program, the viewer can watch a scene and then choose the path they want to follow. Now transferred from tape to DVD, if students choose the wrong path, they get shot. Just like rental DVDs have deleted scenes, featurettes, and alternate endings, more elaborate branching allows viewing shots from alternate camera angles. The viewers can create their own timelines or edit progression, and you can program stops that make the viewer see only what you want them to see. In the pornography business, a DVD allows the viewer to watch different camera angles in a scene. This has been a feature the adult video market has been waiting for.

New Business Ventures

As mentioned before, the process of making or creating a DVD is called authoring. Some people specialize in taking video masters and converting them to DVD, increasing their content, and marketing the final product. One of the reasons DVD excels as a new storage medium is that it takes up less space and is a more permanent archival method than magnetic tape. As I mentioned in a previous chapter, several of my old training interactive videodiscs have been converted to interactive DVDs (and given a new life).

In order to author DVDs, you need two things: a DVD recorder/burner and authoring software. I won't discuss the prices of the DVD recording systems because by the time you read this, new models will have hit the shelves and the prices will have dropped. Let's focus on two authoring systems out there (dozens exist) and how they function. I've used both of these systems and the two have great features. Presently, I've only used the Apple versions of these programs.

As a side note, I received *DVDit* with my scanner. This is a basic authoring program that enables you to create inexpensive DVDs with limited branching. The other two systems, Sonic Solutions DVD Authoring System and DVD Studio Pro, are more expensive but offer more creative freedom in a more secure package. DVD Studio Pro will be discussed in the next section, "How Not to Get Involved."

The Sonic Solutions DVD Authoring System begins with a Sonic DVD Creator audio/video input-output breakout box. Similar to a *nonlinear editing* (NLE) system breakout box, the audio and video leave a video player and the audio enters an audio board, while the video enters via an SDI input and is decoded into an Apple G4 or higher computer. Although the Mac has a DVD Super Drive capable of burning DVDs, the process is far from over. The in and out timecode points are logged for every tape and the process of encoding begins. Similar to capturing in the NLE world, you actually take the audio and video footage and capture it real time onto an external hard drive (a two-hour program takes two hours). The Sonic Creator is not an editor; it's a true DVD Authoring System.

Once all the material is encoded on the external hard drives, the authoring begins. Authoring is basically linking the pictures, subpictures, menus, and titles the DVD contains. Using the authoring program, all the audio and video files, graphics, menus, and FBI warnings are imported.

The authoring process is almost like doing a web page where items are linked to each other. If you have multiple menus, you can bounce back to a larger menu and go to another menu from there and it's seamless. Like designing web pages, it can get very detailed. The software is not something the average person can purchase and just sit down and work with. Even with a web background (it's not a Spiderman thing), it takes a lot of organization, patience, and knowledge. I definitely recommend taking a class or, at the very least, following the tutorial.

Examples Are More Fun

In order to generate interest (and money) for the short film, *The Cryers*, authoring it on DVD was the answer. A prototype DVD would be created first (authored in Sonic) and then mass produced to allow possible investors, rental houses, and festivals to view this masterpiece. In Figure 8-1, we see behind the scenes of shooting for DVD.

The branching begins on paper. You must determine what the viewer will be allowed to see and when. The DVD will be made up of several elements in the branching structure, all accessible from the menu. Like in a restaurant, without a menu you will have a tough time deciding what to order (or watch). *The Cryers* is a 10-minute short video. An additional 35-minute, behind-the-scenes documentary was shot on the making of the video (it's not Ron Howard, but we had to start someplace). Interviews of the cast and crew may also be chosen from the menu as well as deleted scenes and outtakes. For an example of an interactive menu screen, look at Figure 8-2.

Now that the paper path is completed, you can begin constructing the elements of the DVD. After an FBI warning or Interpol disclaimer, the menu is the first thing (you created) that anyone will see. This needs to be interesting because some people start the DVD and then run to the bathroom, get food to eat, or begin doing something else. This "still frame" can be moving, have music playing underneath, and will loop continuously if

Figure 8-1
Behind the scenes on
"The Cryers"

Figure 8-2
Menu screen for the
Cryers DVD

nothing is chosen. The background of this menu screen should be exciting with a still or moving image from the video, clever artwork, or anything that is readable, not too busy, or confusing.

The viewer clicks on the item of choice and immediately travels to that area. In Sonic, the timecode for the start of the selection is entered. With five items on the menu, five timecode in and out points are authored in. After the viewer chooses and watches *The Cryers*, he or she is taken back to the main menu.

If the viewer chooses the outtakes, each outtake must be edited to the next one unless a submenu is generated where the viewer can choose to watch each one individually or all at once. This submenu branches to another submenu where you may select to hear the director's commentary. The most complicated authoring should be broken down into simple menus that ask what is selected, where do I go, and after I watch this, where am I taken to. These simple in and out cues are all the authoring involves (over-simplified, but that's what happens).

At this point, the only costs involved in getting your video to DVD is the cost of the authoring software (rent time on the system or have someone else do it), the time it takes to encode and author (an inexpensive chimp may not work here), and the cost of a blank DVD (cheaper than you think).

Encode Me

Once all the in and out points of the menus are determined, the video should be encoded (digitized). With the timecodes keyed in, the video player will stop and start automatically as it sends the video and audio signals to the hard drive.

When this step is completed, all menus should also be encoded into the system. Everything the DVD needs now resides on the computer's hard drive. At this point, only 2 hours of digital information can fit on a 4.7 gig DVD (more on determining bit real estate later).

Once the branching is tested, the project may be burned onto the DVD. Follow the instructions on your particular recorder/burner and you've just created a DVD. Unlike videos and data CD-ROMs, DVDs, once recorded, are permanent. You cannot record over a branching flaw if you notice it once the disc has been recorded.

Throughout the operation, incorporate a detailed check system. Explicit (not naked) forms should be filled out during the encoding and authoring process. Each menu and navigation route is checked for correct color selection, menu linking, audio and video sync, and so on. I hope creating DVDs will allow me to be recognized as outstanding in my field . . . without having to walk through all that tall grass and getting my shoes wet.

Review: Five Reasons Why DVD Authoring Is the Wave of the Future

1. Unlike videotape, DVDs do not wear out, stretch with age, or end up wrapped around some tree along the interstate. An excellent archival method, each DVD-9 may hold up to five hours of footage in a smaller space.
2. Encoding enables almost any medium to be stored onto a hard drive before being burned onto a DVD. A DVD-R can create a DVD without going into the authoring process if your menus are simple without complicated branching or icons, much like old Jerry Lewis movies from the 1960s.
3. Encoding must be done correctly. If any mistake is made at this level, the final product will be flawed. Check every step of the encoding process to make sure what you see is what you're going to get. No one wants to buy a DVD with six hours of Aunt Edith sunning herself by the pool.
4. Authoring takes patience and organization, and some previous web experience is helpful. The novice should attend classes or begin with simple branching or incur the wrath of the Authoring Police.
5. Older DVD players sometimes have an appetite for new DVDs in that they will not play. Just make sure they have been fed frequently. (You can buy a DVD Snack Mix at your favorite store.)

How Not to Get Involved

If you want to save money in production, one of the best ways is to do something yourself. You can't always do everything, but you should at least know how the process is done so you have a better handle on it.

The presentation of the completed DVD will also add points when showing potential new clients. After spending all the time and expense creating a DVD, don't label the disc with a Sharpie and the outer case with a sticky label. CD Stomper and other disc-labeling programs are inexpensive and make your DVD more presentable and professional. With a color image scanned and pasted on the disc and a front and back jacket insert, you have a DVD that should compete with the pros (at least from the labeling perspective).

In DVD production and authoring, doing the work yourself will save you cash, but if that's the way you want to handle it, training is the best answer.

In a Class by Itself

Now that DVDs are here to stay, it's important to know how to create your own from scratch. Wanting to learn more about Apple's DVD Studio Pro 1.5, five experienced professionals and I enrolled in a two-day course that would enlighten as well as entertain.

Coming into the class with different levels of experience, some wanted to demystify the process of authoring, others desired to learn the nuances of scripting, while a few needed guidance encoding their MPEG-2 video into an interactive form.

My humble advice to anyone desiring to branch into DVD authoring is to take a course on the software. DVD Studio Pro isn't a difficult program, but the manual leaves a video professional like me slightly lost when new terms and procedures are described. Buying one of the few books available on DVD authoring may be helpful, but nothing beats hands-on training with an experienced instructor.

That brings up another imperative item. If you enroll in any DVD authoring program, research the credentials of the instructor. I could easily buy a book on authoring and attempt to train other poor souls in the eccentricities of authoring (not you, other poor souls), but as soon as the first question was asked, I would sound like Ralph Kramden once Alice discovers his scheme.

Free Is In

Get on the Internet and download all the free stuff you can on DVD authoring. In class, we were instructed to do so, and Apple, DVD Studio Pro, and 2-Pop all offer free information to make your authoring process that much better.

When planning any DVD project, bit space (or real estate) and navigation (menus) are two factors that must be considered. Some first-time authors try to cram too much information in a 4.7-gig DVD. A downloadable "bit-budgeting program" is a detailed spreadsheet that interactively determines how much data real estate at a predetermined encoding video rate (video, audio, and surround sound) will actually fit on a high-quality (9.8 *megabits per second* [Mbps]) DVD-5. It's best to find out early if too much information is expected to fit onto your DVD. All current programs will keep on encoding until full and then suddenly stop. With bit budgeting, you'll know ahead of time to make the necessary revisions (a much better way to fly).

Menus and their design enable the end user to travel through the branching roadmap to the various programs on the DVD. DVD Studio Pro displays an item in *italics* when something critical is missing. In my ancient days of interactive videodisc branching, intricate and complicated flow charts had to be designed and implemented as part of the scripting process. Studio Pro leaps light years ahead and uses an extremely easy to use scripting tool.

Classes also alert the student to keyboard shortcuts and tips that user manuals often fail to mention. I learned far more from the class and it actually helped me to understand the book better.

Your imagination is the only thing limiting you in a software program like this. The most complicated and intricate Hollywood DVD may easily be broken down and created on your own system. Tools like Photoshop and Illustrator are useful in designing functional menus. The drag-and-drop technology of DVD Studio is easier (in my opinion) than Sonic. With arrows connecting each menu item, it's almost impossible not to know where your markers are (markers being points the user is branched to).

Like a detailed word-processing program, files in DVD Studio Pro are created for each step of the branching process. With a little desktop management skill, most could easily have the correct video playing at the predetermined time.

Taking an inexpensive two-day course in the authoring program will pay for itself after your first DVD burning. The expense of the knowledge is minimal. I now feel comfortable creating an interactive menu as complicated as any Hollywood megabuck production (and I'm slow).

Review: Five Steps to Successful DVD Authoring

1. Once you purchase a DVD authoring program, it is critical that you take the time to attend a course. Books are helpful, but you can't ask a book any questions, unless the book walks and talks and then you have another problem on your hands.

(continued)

(continued)

2. Make sure the instructor is knowledgeable on the software. You should be able to ask any question without stumping them. If your instructor is constantly looking at his or her shirtsleeve for helpful hints, you may know more than he or she does.

3. Even if you are proficient in a particular program, the new knowledge gained in a seminar will speed your DVD authoring process. When someone has been authoring and discusses shortcuts to the class, you will learn from their mistakes.

4. Don't be afraid to ask questions in class. Your individual problem may be the same issue that someone else has overcome. The only stupid question is the one that isn't asked.

5. Use all the information gained in the course and apply it to your DVD design. Download manuals, updates, and auxiliary items from web sites and be willing to try new things, except that new flavor of Slurpy the mini-mart offers.

Reaching Millions by Spending Thousands

The adage that you have to spend money to make money is true in the world of DVD. But you don't have to spend that much to reach people. Since everyone has access to DVD players, presenting one to someone is an open invitation to view it. Like I mentioned earlier, if someone gives you a DVD of anything, you will watch it because it's new, nifty technology. If you received a new book, unless it were something you were really interested in (like this book), it would soon be gathering dust.

How It's Done

With a DVD burner, you could make 1,000 copies to distribute your project to the world, but that would take a lot of time, effort, and expense. The only way to mass duplicate a DVD or Video CD (presently) is to have a professional resource do it. Larger companies have the duplication and pressing equipment, the expertise, and the personnel to make numerous copies. You can pave the way by creating the authoring and master copy, but leave the replication (or clones) to the pros.

A colleague of mine, Todd Taylor, creates and produces hundreds of thousands of DVDs per year. He doesn't replicate them in his basement and stack them in trucks for midnight shipments. Instead he takes the preproduced video, encodes, authors, and checks it before sending his master to a DLT tape. The DLT is essentially a first copy of the DVD from which a mastering facility will create a check disc. Any DVD authoring program will get you to this stage. It would take months of constant toiling to replicate a 60-minute DVD 1,000 times, with each replication happening in real time (plus finalization).

The DLT offers numerous benefits. All the digital information (menus, video, audio, branching, graphics, and scripting) are dumped into the DLT. This is sent to the mastering facility and a check disc or proof disc is created. This one copy is an exact duplicate of the DVD and every reproduced copy will look exactly like it.

The check disc is a time- and cost-saving way of double-checking your work before the expense of thousands of DVDs are pressed. If you find an error in the video, menus, or branching now, you can correct it and have another check disc made instead of recalling every duplication. Ideally, you should double-check all your work before the DLT stage; most programs offer this feature to troubleshoot your disc before the mastering phase.

Once the check disc has been approved, make as many copies as you like. You now have a mass-produced DVD that will play in every machine and computer on the planet (something a DVD-R won't do). The glass master created from your DLT is used as a stamper to create numerous replications instead of burning each and every copy. This is a more accurate way of cloning than copying and burning. A dandy way of looking at a DVD and telling if it is mass produced or a one-time recordable is by the color:

- **Blue cast**   A DVD-R or DVD-RAM (recordable)
- **Silver cast**   A single-layer, mass-duplicated DVD
- **Gold cast**   A dual-layer, mass-duplicated DVD

How Much Is It Going to Cost Me?

This has been the burning question throughout the ages. Everyone wants to get the most for the least amount of money, and I'm no different.

I produced a documentary for a couple whose 15-year-old daughter had been wrongfully killed. Given too much medication in the hospital, she died in her sleep, leaving only memories for her parents. I videotaped them in interview fashion and compiled a 20-minute video with their reminiscence of her intercut with home videos and photos. It's not important in this case how I created the program, but how this video could be seen by others.

I burned a DVD-R copy, scanned a photograph for the jacket and disc surface. This "one-run" copy would not play in a DVD player older than two years. The girl's father asked me if he could have copies to give to family and friends as a memorial to his daughter. The burning question—how much?

I told him I could duplicate the program on VHS from the digital DVCAM master at a cost of roughly $1.50 per tape. This would include duplication, tape stock, and labeling. Depending on the number of copies, his costs could go down. But with VHS, the tapes could get overplayed, break, stretch, snap, or get erased by a copy sitting too close to a magnet.

He then asked about DVD. With an initial run of 100 copies, would it be cost effective to replicate on DVD-R or DLT tape? This wasn't a question to ask him, so I did some research.

Since only 20 minutes of moving video appeared on the program, no menu screens or branching were needed, and the program would easily fit on one side of a DVD. This could be burned on a DVD-R. By loading one blank DVD-R into the burner at around $4, 35 minutes to record each disc with finalizing, labeling, and jacket preparation, the end cost with labor was almost $40 per copy—final cost $4,000.

Labs in New York and Los Angeles would duplicate 100 DVD-Rs at a cost of $12 a piece (2002 prices), plus shipping, jewel cases, and labeling. These are the same quality I could do, but they have more than one machine—final cost $1,200 plus. I would have to record one at a time 100 times.

If we mass produced it, a DLT would still be created even though no branching was involved. The cost to make the check disc was $900 (in this case since no branching is needed a check disc wasn't necessary). If we upped the quantity to 1,000 (the minimum for glass mastering), the cost is $1.15 each plus shipping, jewel cases, and labeling—final cost $1,150 plus. This is a much better deal. For $50 less, he has 900 more copies that will play anywhere where as the 100 DVD-Rs might not play in some machines. We could give him his 100 copies, but I doubt he has a need for the other 900, so who pays for this? It sounds better than $4,000 but still is steep.

At this stage of life, it's not cost effective to replicate less than 1,000 copies of your program if you need DVD as the medium. If my same project were going to rental stores, for purchase at events, or any other method of distribution, 1,000 would be a great starting point. Currently, increasing that number to 1,500 lowers the cost per disc to less than $1—far cheaper than VHS and more permanent.

I just completed a video for a national travel agency that wants the world to see how beautiful their tropical resort really is. If they want that beauty on DVD, they need more than 1,000 copies. No one wants to get a DVD that won't play in their system and if you duplicated less than 1,000 on DVD-R, 25 percent of them would not be compatible with the players.

My advice is to reach for the millions and make more than 1,000 copies if at all possible. The cheapest Hollywood DVD in stores is $5 and those are old movies. Your masterpiece could be sold for as little as $3 and still make over 200 percent profit. Those are great numbers in anyone's book.

Features

Creating Features on Digital Video

Feature films have always had a stigma attached to them. To some, the word feature brings to mind name Hollywood stars, huge budgets, and saturation advertising. As a video person, you've read about the small independents trying to compete with the giants. Every once in a while, a great independent film will surface and actually do exceedingly well. This chapter discusses making those types of features on a shoestring budget.

Numerous features are made with a great idea, a bunch of people, and a digital camera. Today's consumer three-chip cameras compete with professional units in image output; it just takes talent behind the lens to pull it off. People don't mind watching a video feature. The "film look" isn't expected as it once was before. As long as viewers can pop the tape or DVD into their home theater, they are entertained. The need no longer exists to transfer the video to film, edit it, and project the print in the cinema. Even Disney makes direct-to video releases that never see the light of day at the megaplex.

The Secret Ingredient

You've seen hundreds of features and know basically what it takes to make them: a good story, convincing acting, and, most importantly, money. Without having deep pockets, most independent features never see the light of day. I'd like to break that myth right now. If you have a story, crew, and actors, money shouldn't be what's holding you back.

Numerous features have been shot on digital video. With three-chip cameras costing a few thousand, the most expensive obstacle has passed. With tape stock being far cheaper than film, processing, and printing, independents have made features for only a few thousand. The only way to make this a reality is in the preproduction and planning stage. Without this step, you may be doomed to the worst that could happen—an uncompleted project. Unless the project is finished, you will never recoup any of your invested money. I'll discuss marketing your completed project in the next section.

One Step at a Time

A feature is just a long video project. Instead of running 30 minutes, it may run three times that long. Shooting a feature is no more difficult; doing a great job on a short may lead to this longer format product. If you look at it that way instead about worrying how difficult features are to make, the job will be easier.

By breaking a feature (a project that runs about 90 minutes or longer) into smaller steps (like doing several short videos), the task seems less daunting. Once you have the script, take the time to break it down into a storyboard and a shooting script. At this stage of the game you will have a better idea of what you're getting into.

Some will say don't bite off more than you can chew by shooting a science fiction or action-laden epic. Obviously, the more complicated the script, the more money will be spent in various areas. But the independents have something that Hollywood will never have—freedom. With massive overhead, salaries, and thousands of personnel to keep busy, the studios have to spend a minimum of millions per picture just to stay in business. The independents don't have that burden.

A script that focuses on the story with minimal locations and a smaller cast is less expensive to produce. My first self-produced feature was just that; I tried to save money in every phase.

It Almost Worked

I wrote my first feature (which saved money) and I kept the budget in mind while doing so. This is helpful if at all possible. The story involved a young couple inheriting a mansion in Vermont and discovering it was haunted. The comedy started with the couple trying to get to the house with hundreds of misadventures along the way. Once at the house, the parties trying to scare them away had learned too much from Vaudeville (this lessened the expense of special effects because they were supposed to look hokey and phony).

The cast only involved eight people (not too many mouths to feed), one-third of the story took place in a car (an inexpensive rolling set), and the rest of the video was in the mansion (a borrowed home that was for sale).

As I mentioned earlier, break the script down into locations, setups, and the number of shooting days. The biggest mistake independents make is to cram too much into too little day. I originally figured 20 12-hour days to complete the shooting; this would kill my cast and crew and also make them mad at me. Instead, the product took 30 10-hour days—much less grueling.

Deferred salaries or a copy of the finished video helped save money, but I was expecting people to put in unrealistically long days to save a few bucks. Whatever amount of time you believe the project will take, add at least 25 percent. In low-budget shoots, effects don't work when planned. Cast and crew show up late, take long breaks, and may have other money-making commitments (jobs). Since you're trying to save money, everyone and everything seems to be against you. Make the shooting schedule longer than you anticipate and your crew will be happier campers. If you finish early, you may even get brownie points.

Once everything has been scheduled in detail, secure the locations. Free is the word of the day. For the exposure (there's that film term again), most will allow their possession to be videotaped. You need to compensate *everyone* in some manner. It may not have to be money, but arrange to trade, swap, or barter for the use of what they have. I happened to notice a Vermont mansion while on vacation. The home was for sale and furnished, and it had the gothic appeal I was looking for.

When I asked the agent about the house, he immediately said no. He expected thousands of people with trailers and massive lights storming the edifice. When I pleaded my case with the vacated owners, they approved. It sometimes takes talking to the right people to get what you want. Don't pretend you're a hot-shot Hollywood producer because they will expect that type of compensation. The owners were flattered, and as long as we didn't destroy anything (and pay for it if we did) they were fine with it. My point is don't be afraid to ask for anything—you just may get it (probably not, but it's better to be positive). Video is a magic word; whenever you speak it, people are guaranteed to listen.

Figure 9-1 shows an example of what a video feature set looks like (no different than any other video shoot). The mansion in question can be seen in Figure 9-2. It was being used as a wallpapering school.

Try as much location shooting as possible. These places already exist and you don't need the expensive of building sets. All our locations and cars were obtained for little or no money; it just takes time to find them. More than two months were spent trying to line everything up for the shoot. Talk to your local chamber of commerce, film commissions, and anyone in local government about your project. This PR helps the community as well as you by getting the word out. The local newspaper will also offer free publicity in the form of photos and a story.

People are fascinated when they read someone is shooting a "movie" in their home town. If they see someone with a camera, they immediately believe Hollywood has found them—use that to your advantage. This is also a dandy way to get free extras. Let the property owner be a nonspeaking extra in your feature; that alone will get the use of their possession. Adding digital extras is too expensive and doesn't allow you to use such an option.

Figure 9-1
On the set of a video
feature

Figure 9-2
The Vermont
mansion

Finding People

The cast and crew are the next most important element once the script is completed. A top-notch talented crew will make any video sing. The director, camera person, and sound person are key personnel that will make an independent film not look like it is. Don't worry about having nine producers or executive producers; these people don't do that much and they will just suck up the budget.

One of the potential investors in my feature wanted a producer credit because he put up some money. I granted him the title as long as he didn't get in the way; he did on every single shot. I ended up thanking him for his time and returned his pittance; it was not worth the aggravation. If you make the same sort of promise, be careful what you may be getting into. This clown actually slowed the production down and cost me crew time.

You can never have too few *production assistants* (PAs). Schools, interns, or clubs will offer a multitude of these excited individuals who will keep things flowing on a shoot. These people are more helpful on shoots than producers because they do the real work—getting the nearly impossible done. You will attract attention everywhere you travel with a camera. Sometimes your PAs will have to deal with these crowds.

The World Is Not a Stage

The actors will also make the difference in front of the camera. I needed a sixtyish man for the bad guy lead. I could have obtained a 20 year old at no charge (for the video), but the make-up would never have been convincing. If you need someone with a certain look or age, try to find them. This adds believability and creditability to your independent video. Advertise everywhere and don't pretend you're a big shot. This video could lead to bigger things for everyone and make that point known. If you pretend you're something you're not, it will explode.

I was able to secure real police cars just by asking and explaining in detail what I wanted. The early material in this book explains other ways of getting what you need. The same advice holds true whether shooting 30 seconds or 90 minutes. Once again, don't be afraid to try anything; all they can do is shoot you.

You will have more control on a sound stage than on location, but the cost and time savings of locations are worth the possible audio dubbing and waiting for the men wearing black suits in the van to leave.

The Best You Can't Afford

Your production values have to exceed all expectations. You will be competing with the pros and your video should look like theirs. Get the best camera you can afford (to rent or purchase), light to the best of your ability, and mike everything. Shaky cameras, underexposed images, and muffled sound all point to a poorly planned production. You will have to cut corners at times, but the viewer should never be allowed to see that.

In my video, the actors were believable in their parts, they rehearsed often before the cameras rolled, and at no time did anyone walk off in a huff. Don't be stingy in a low budget. A plate of bagels, coffee, or some potato chips on the set at all times makes the crew and cast happier. It doesn't cost much for food, but without it things go slower.

If you are set to wrap at a specific time, don't exceed that unless you are left with no other options. People working for little or no money get just as tired as well-paid folks. Make the video a team effort. If the sound person has a great idea, use it. Everyone has as much to gain or lose as you do.

I mentioned it earlier and it is worth repeating: A feature is no more difficult than any other narrative video; it is only longer. With that in mind, there's nothing stopping you from shooting the next *Gone With the Wind* (except that the original actors are dead).

Whatever It Takes to Get the Shot

This heading should always be your motto. I have been in a lot of unusual places and was always trying to hang by my tail holding the camera in my teeth. Figure 9-3 shows a unique way to get a steady subjective shot of a car traveling toward you. By sitting in the trunk with a tripod-mounted camera, the car's shocks will dampen any bounce. Just make sure no one closes it while you're still inside.

Figure 9-4 is at the other end of the spectrum. If you want to shoot something with water (a lake) in the foreground, the no-budget approach is to wade out into the liquid and shoot it from there (just don't get anything wet).

Figure 9-3
The trunk point of
view shot

Figure 9-4
The lake and you

Review: Thirteen Ways to be Featured in a Feature

1. Break any long-length video into smaller, more manageable sections. Nobody sets out to shoot an entire feature all at once; each segment is handled one at a time. Since you have experience in shorter-length productions, a feature-length project is now more manageable. The amount of tranquilizers you need will also be less.

2. The script will actually pull more weight than the visuals. If your script is weak, take the time it needs to make it better before shooting begins. The public will remember a great story even if the camera work is shaky. People have said they planned to use shaky cameras, but no one ever wrote a bad script on purpose. Doctors call prescriptions for scripts also, and I could tell you about the time I added a second floor to a building, but that's another story (great third-grade joke).

3. Don't try to kill your cast and crew by making them work excessively long days. You probably aren't paying them a fortune so it's better to have them like you through most of the production.

4. If money is an issue (is it?), defer salaries or offer a percentage of the profits. Anyone working on your feature needs to be compensated in some way. The experience is invaluable, but five minutes alone with a Ferrari is worth much more.

5. Whatever your proposed budget happens to be, add 25 percent. The shooting will take longer than expected, unplanned expenses crop up, and you never know what's going to happen along the way to slow you down. If you plan for it up front and don't need it (you most likely will), throw a party and buy the potato chips in the cans rather than the warehouse club-sized bags.

6. Have everything detailed: script breakdown, call sheet, camera blocking, and so on. You can never overplan a video production. The more detail that goes into the preparation, the fewer mistakes will occur when shooting begins. You will notice problems when planning and this is the place to solve them, not when you are paying for a full cast and crew. Most horror stories come from poor planning, except *Nightmare on Elm Street,* which came from a good writer.

7. Try to secure free locations by explaining your plight, lack of budget, and high enthusiasm. It never hurts to ask and you may even get more than you bargained for. Make sure you know all that's involved with the word "free." I once got a free location, but I had to use the owner's talentless, toothless, bald, and mangy cocker spaniel as the lead.

8. When trying to obtain no-cost services while planning a feature, don't pretend you're a big-shot producer. If you do, people will expect to get paid much more. This is one instance where less is more (the other is when sharing a bagel with a loved one).

9. Get the best cast, equipment, and crew you can afford, beg, borrow, or steal. This can be a learning experience, but skimping in any area may show up in the final result. I should have known that the cheap, two-legged tripod wouldn't work out.

(continued)

10. The more PAs running around the better. You will have too many other headaches during the shoot; let a PA handle the day-to-day routine. What happens when your father, who lives in Pennsylvania, becomes a PA? I guess that would be a Pa from PA who is a PA.
11. Get actors of the correct age to fill the part. Make-up gets too expensive making a 70 year old to look like 23 or vice versa. These people are available; you just have to find them. Look in the Yellow Pages under . . .
12. End each day at the predetermined time if at all possible. Taking advantage of people by abusing them will bite you in the end. Some things are out of your control, but keep an eye on excessively long days. Those will come later in the delivery room.

Marketing a Long Movie

Selling the world a 90-minute video is no different than selling them a 10-minute piece—only easier. Your finished feature has hundreds of outlets that might not be possible with a shorter subject. It seems people have more time to watch something that's 90 minutes long rather than 10. Maybe it's because they won't have to be entertained again for an hour and a half. So who do you tell when you have finished your digital video feature—everyone.

Going to the Market

If no one ever sees what you just completed (the video), it was only a learning experience. But if you want to recoup all $300 you spent on tape stock, people must see it. I'll explain this in a little more detail in the next section, but you must advertise before, during, and after the video. A movie poster works well and is inexpensive to create, a newspaper may advertise a public showing, DVD copies for sale and rent in stores will get your movie in people's homes, and word of mouth will have everyone talking about you (they do already and most of it is nice). Marketing your video can be done in two ways: getting a distribution deal or doing it yourself. Let's begin with the poor person's approach—I'll do it on my own.

The Movie Poster

Everyone has stopped and looked at a movie poster. They are colorful, show an exciting action sequence, and display your name in front of the audience. The best time to do a movie (or video) poster advertising your project is well before it's completed. The old joke about revealing one phrase at a time promoting a new picture might work here ("It's Coming . . . ").

Your goal is to excite, stimulate, and titillate the viewer by displaying a poster anywhere and everywhere someone may notice it. With programs like Photoshop and Illustrator, you can create a 20 × 30 advertisement as well as smaller 8 × 10 copies for store windows. The most important item to mention in the poster is that it stars or features local talent. People will want to see themselves and others in a local, professional production. Figure 9-5 showcases my first horror epic in a movie poster format.

I worked on the lowest-grossing children's feature film to date when it was shot in my hometown. At a cost of $50 million, the film made less than $1 million in its first weekend; that is downright horrible for a children's movie. But it did phenomenally well locally because everyone wanted to see themselves onscreen and in the credits. Another big-budget film in the

Figure 9-5
Nobody movie poster

Figure 9-6
A feature-sized crew

1960s, *The Molly Maguires*, shot in my childhood town, did poorly nationally, but not locally.

At this stage, begin locally marketing your feature and then move on to nationally. You know people will watch it where you live, but out of town may be another story. At least you'll recoup some of the expenses before you travel with your video. Because we're still in the "I'll do it myself" section, I did my poster all alone (except that Staples printed the copies). Figure 9-6 shows the crew size on a large-budget feature. (See if you can find me. Hint: I took the picture.)

Newspaper Advertisement

My father worked as an assistant editor in a newspaper for more than 40 years. During that time I got free advertisements anytime I wanted the town to preview my exciting epics. He even saved me money by printing up dummy newspaper front pages that I used in all my productions. When people want to be entertained, they look in the newspaper to find what's happening in town.

Try to find an inexpensive (free) place to show your work via projection. Moose, Lion, and Mason Halls will allow local video projections from time to time. Like Andy Hardy when he wanted to put on a show, ask around to see where you can find suitable accommodations.

This brings up an issue. How many people do you expect to attend your gala premiere? I've borrowed a movie theater after hours in the hopes that 1,000 people would attend, and I didn't fill the place (the first row was pretty crowded).

Unless you ask for RSVP, you will have no clue as to the size of the crowd. You can always try to drum up more business by driving around town shouting that your video is "showing at the Roxie at 8." If you charge little or no admission, you will have a larger audience, as will a wine and cheese enticement (a larger expense). A hundred enticements can be made to get people into your screening. See what others have done and what worked for them. If you book a small place and "sell out," that's great. It might be better to spend less and have a more intimate showing and have second, third, and fourth screenings than barely paying the cost of heating the large, vacant auditorium.

Your local church would also be another avenue to explore. Most churches have better video projection and sound systems than civic organizations. In addition, send out invitations to your screenings: Invite the media, local government, and friends. The more people asked, the potentially larger the audience. Schools and colleges are also set up for screenings, so talk to people who work there.

On the day (or night) of the screening, play it up like Hollywood (now it's okay to do it). Bathe, wear your best clothing, and greet guests as they arrive. Have friends wave flashlights outside to attract attention, get people with the longest cars to pull up outside, and ask a local anchor (the TV kind) to be an emcee. Let your imagination run wild and play this showing up like it's the biggest event of the year. Use adjectives like amazing, stupendous, and colossal (I'm kidding, but you get the idea). This is your world premiere and if your video makes it there, it can make it anywhere (everybody sing).

To DVD or Not to DVD

After your screening has been a success and you have DVD copies for sale afterward, it's time to hit the rest of the community. Video stores like Blockbuster, Hollywood Video, and numerous other chains all have outlets in your area. It's best to start with these local stores because the manager in the store has some pull. He or she may have to check with corporate, but with them on your side you will have an easier time making that a reality.

I've seen local video features for rent in my nearest Blockbuster. Under the section, "New Releases," certain videos and DVDs had a computer-printed sign saying the feature was locally shot and produced. At least, someone in the store will pick up the box and look at it (we'll get into packaging in the next section). If the video appeals to them, they will rent it, but at the very least they picked it up.

With at least 10 copies per store in our town, that would mean 100 copies would be available for rent. This is another good place for your poster, in the window of the video store. It also helps if the sales people play your video on the overhead monitors (saturation advertising).

With the price of DVD production being low, it's cost effective to make at least 1,500 DVDs on your first run. When a video retailer gets 10 DVDs for rental, they buy them at a bulk cost (let's say $15 a piece). The store will then rent the piece for at least one year with DVDs having an almost unlimited playing period without damage. At the same prices, that's $150 per store with them charging $4 or more per rental. They will recoup their money and you have almost paid for the cost of 1,500 copies by selling them to 10 outlets.

Each of these video retailers offers DVDs for purchase. Ask if they will take 10 copies each on consignment (they won't pay for them unless they sell it). Selling at $19.95 nets you even more profit. Now each of the 10 stores has 20 DVDs; you only have 1,300 left. I didn't even mention VHS. Some rental customers still look for a tape copy rather than its digital brother.

The next outlet is where people buy DVDs and tapes: Wal-Mart, Kmart, Circuit City, and so on. Once again, start local by talking to the store managers and see what you can come up with. Your video feature will start getting into the hands of the public. Follow this route and hit every retailer in your area (the PAs can help here also once you have set up the account).

When to Have a Big Mouth

Have every actor and crew member have copies of your feature at their fingertips. Each person will tell others about the video and happen to have a copy with them. Whether the copies are loaned, bartered, or sold, it's up to you. This free form of advertising lets people know about the video and may peak their curiosity.

I've even gone as far as showing my finished video at parties where people will wander by and happen to notice it playing. This is sleazy, but I was invited to do so (play the video, not be sleazy).

Some people don't feel comfortable with this sales approach and if you'd rather have someone else do the selling, it may be time to relinquish a little of your profits.

Let's Make a Distribution Deal

If I wanted any of my videos to make money, I needed to find a distributor. A distributor will handle the duplication, label printing, and the all-important marketing at their own expense—at a price to you. For a percentage, usually 80 to 85 percent, they will take care of all the headaches leaving you with 15 to 20 percent of the profits. That may not sound like much, but they will have access to markets you won't. If you produce the video in northern Oregon, how are people in southern Florida going to hear about it? You can always do the Loretta Lynn approach (the movie *Sweet Dreams*) and drive from county to county, but the distributors do this all the time. They are experts at it and can save you a lot of driving. You will make more money in the long run, but they will take the largest bite because they now have the biggest expense.

It's much like you reading this book. If I had peddled each copy to every bookstore and web outlet, I'd have more money (to fund other videos), but I'd have to pay for printing, packaging, shipping, and advertising.

Like any contractual agreement, know what you are getting into before signing on the dotted line. Who has final control? What about overseas marketing? Make sure you know your rights before you sign them away. For further information, Michael Weise has a great book on distributing an independent video; The Independent Film & Video Makers Guide type his name into any search engine.

The next section will discuss how to effectively advertise your video. Read on because I won't tell you how it's going to end.

Review: Eight Nifty Ways to Market Your Video

1. Make a movie poster detailing your production. You've all seen these in movie theatre lobbies, stores, and billboards. This is cheap advertising, but make sure you ask before just sticking up a poster. "I'm really sorry lady. I didn't know that was your . . ."

(continued)

(continued)

2. Market your project locally. Since you shot it in your town, the public needs to know that fact. Even if the project flops elsewhere, it will do well in your hometown. That's mainly because your parents and grandparents will buy all the tickets to make you feel good.

3. Advertise in the newspaper and radio. During production, a story about the video helps, but once it's completed, the world has to know. By playing up the story and what's involved in the video, people will be clamoring to watch it. Once you start the video, lock the doors.

4. Try to arrange a free screening location. This can be taken two ways. The place where your video is shown should definitely be free if possible, but should you charge admission? You will get more people and better word of mouth advertising if the screening is at no cost. However, charging to watch the piece is one way of recouping a little of your budget. This should depend on your situation and how big Lumpy really looks, who has been following you around all week.

5. Make VHS and DVD copies. A lot of people don't like viewing videos in large crowded places. Some prefer to watch something in the privacy of their home. Copies can be made available after the screening or purchased from video pirates down the street.

6. Hit the local video retailers and rental stores with your movie. People will want to see something that was made in their town. Make sure you hit every place that would possibly sell or rent your epic. This is one place where those coin-operated booths aren't a good idea.

7. Have your cast and crew advertise. This word of mouth or word of T-shirt and baseball cap will get the public interested. Just make sure everything is spelled correctly. A woman displaying the word "Grip" across her chest causes problems.

8. Carry copies of the project with you wherever you go. You may never know where you might pick up a potential sale (unless your pimp gives you the night off).

Advertising—Inexpensively

Look at all the fun Darren Stevens had during the run of "Bewitched." He was an advertising executive, he always made the client happy with a great concept, he sometimes caught Larry Tate off guard, and he was constantly trying to keep Endora from changing him into a turnip. Often without resorting to magic, he came up with a clever catch phrase that made people want his client's product. You can also do the same without magic if you add one thing Darren didn't—great packaging.

Everything associated with your video is considered packaging. How the video box or case looks; your advertising to the world by print, audio, and video; and how the feature can get into the hands of the people are all part of the "package."

Coming Attractions

Although your video isn't necessarily a Hollywood blockbuster, do what they do to showcase your project—make a trailer (not the thing you tow behind your Camry). The first thing anyone hears about a new movie is the advertising. A television, magazine, or newspaper advertisement is fine, but expensive. If you know people in these markets, ask them for a small (free?) blurb in their periodical or show. The only wasted lead was the one that wasn't asked. The other approach that has worked for years is the preview trailer.

The trailer is a neatly cut 30- to 60-second version of your feature. This fast-paced trailer may tease (only show so much but not too much), build suspense (showing the best and most expensive scenes), create laughter (the funniest bits are strung together), cause character identification (the viewer will relate to what the lead is going through), and showcase your music (the soundtrack of the trailer must pull the viewer along).

Any of these approaches will be a successful tool to explain your feature video in an abbreviated amount of time. How many times have you seen a preview but hated the picture? That's because they used the best bits and great cutting to get people to see the movie (it worked) because that's all they had going for it. You don't want to trick people into believing your epic is what it's not, but if the trailer is boring, you've lost your audience. I've rarely seen a boring trailer. Even a film I have no interest in seeing can entice me with a good trailer.

I fondly remember the trailer for *Baby Geniuses* with Kathleen Turner. I would watch Ms. Turner doing *anything* (my wife knows about this, we've "talked"), but somehow this concept did not hold my interest. Through digital manipulation, the baby's mouth would move and an actor would supply the dialog. The trailer showed a thousand examples of this with Kathleen Turner (did I mention her before?) and immediately the viewer knew what the premise of the movie was about. I knew from the trailer that the movie would not hold my concentration after *she* left the shot. Now on TV, the one-star film (the critics were generous) is shown often to fill the void. This was one example of a trailer that showed too much, making the viewer lose interest.

I've seen trailers that should have won the Academy Award, whereas the film itself left me cold. In order to save money (that's me; what can I say?), I wrote, edited, scored, narrated, and played the music in the trailer for my horror film, *Nobody*.

Nobody was a suspenseful tale about a disturbed woman who murders her boyfriend who doesn't want to stay dead—or so it appears. Is he really chasing her or is it all in her mind? This wasn't great scripting, but it made people think. The writing was easy. I edited together the best parts in a montage but never showed who or what was tracking her. The soundtrack was a scary piano tune much like the one in John Carpenter's *Halloween*. The only trouble was I didn't know how to play the piano. By simply pressing down on a piano key, a sound emanates and by following through with a sequence of keys—a song is born.

Once your trailer has been completed, it has to be seen. Wherever you set up to show your video, the trailer should be there too. Movie theaters won't play your video trailer unless set up to do so, but video rental stores, Wal-Mart-type stores, or any other market that might benefit from the showing of your video will work. The object is to have it made; people will always spare a half a minute to watch something.

Catching the Wave

Although rarely mentioned on *Bewitched*, a package that gets people's attention will draw their eye to your video more than a catchy slogan. While your video is being duplicated in DVD or VHS, the packaging concept should be at the printers. You are more than welcome to design your own copy, artwork, and inserts, but making more than 1,000 images takes a lot of time, equipment, and expense.

Saving money in this area is easy. The first step is to design the label for the DVD or video cassette. A full-color image is better than black and white, silk screening is more expensive (but looks nicer) than a label adhered to the package, and triple-check that everything is spelled corrected. Mistakes scream low budget and amateur, and there is no sense in giving that away.

The least expensive route is to buy a CD Stomper or Avery label package and, using your imagination, photographs, and artwork, design all the labeling. You can print everything out on a color printer and if less than 1,000 images are needed, you might save some cash but put a strain on your equipment. Don't skimp in this area. Do the best you can afford to and move onto the next step.

Creating a Stir

Now that your video, trailer, and packaging are completed, you need the public stirred up into frenzy about the new release. Anything that will attract attention (legally) is one way to start. What gets you to stop what you're doing and look and listen (besides a gun)?

Someone talking loudly about something will make you stop but also offends. Radio may be your ally in this arena. Disc jockeys are always giving away albums, tickets, or something to get people's attention. That's what you need. Ask if a radio personality will pitch your video if not on the air, at least locally when on a remote. Give them a few copies and plead with them to exercise their golden vocal cords and help a local person out.

If blaring Monkees music isn't your bag, ask the owner of an establishment if you can set up a TV monitor and show your video or trailer. Compensate them in some way, but have drawings for free things. People will always stop when they have a chance of getting something for nothing. Why do you think people at trade shows always have little doodads lying around? These sparkly objects cause the magpies in the crowd to stop and look. The lure has been set and they'll take the bait. This is the only way to get someone to sign up for something—if they have a chance of getting something for nothing.

I've even gone to the expense of having pens, pencils, mugs, and key chains printed out with the name of my video on it. This will cost a few dollars, 12 year olds will grab them by the handful, but people will take a key chain, pen, or mug no matter what slogan is on it . . . because it's free. Before they run off, at least point to your video screen. They may think it's a hidden camera and see if they can see themselves.

It may not be the easiest thing to set something like this up all by yourself, but piggyback with another fundraising attempt, trade show, or fair. I've had the audacity to set up next to a booth where they were giving away a new car. I asked if I could, of course, but a car will draw a larger crowd that anything else. I told the barker that my monitor and video would keep people occupied while standing in line waiting to sign their name for the car. I didn't sell many videos, but I did create a stir.

You're creative; you're in video—use that to your advantage. Look what Lucy and Ethel did to attract attention. Tone that down a bit and follow their lead. You don't want Ricky to get mad at you.

The Press Junket

I've been to more than my share of these events. The star will travel the country to pitch his or her new film, answering questions by the press. Why not have a local press junket with your stars?

Rent a DVD and look at some of the special features. *Vanilla Sky* with Tom Cruise included the press junket as part of the DVD. This is an example of how it's done. Tom won't help you with yours, but you have an example to follow.

Local TV stations have talk shows during the day; this is one avenue to explore. Your leads will get on TV, the interviewer will ask questions about the video, and you've just gained a free television advertisement.

Take this one step further and invite the media (newspapers and TV) to a hotel lobby and allow them to question (not interrogate) your cast. These free public relations will go farther than you think. Even the other guests at the hotel will see the cameras and lights and wonder what's going on.

In my sleazier days, I've set up a camera with no tape, lights, and my actors and staged a phony press junket. Doing this in a public place attracts attention and it also helped with my free advertising.

The bottom line is to do whatever you can to spread the word about your video. If it's inexpensive, you'll feel that more proud of yourself.

Review: Five Ways to Advertise While Wearing a Shirt

1. Make a trailer of your production. Don't lie or show something that isn't in your video, but condense the story in a 30-second piece. This blurb can be shown anywhere and will attract attention. Of course, you can attract attention too, but let's not go there.

2. Create VHS and DVD packaging that will catch the eye. Colorful and not too wordy are the key concepts. Cheaply made labels will make a low-budget production look just that. Have you ever picked up a DVD package and it just looked cheap? Don't let this happen to you: Buy war bonds, save the Alamo, and begin the beguine.

3. Get your epic seen—anywhere and everywhere. It may make you sick seeing it ten thousand times, but the viewing audience needs a saturation campaign. Anytime your trailer is shown, that may generate one additional sale (like when they had sweat socks on sale at three pairs for $1).

4. Merchandising works for the big guys as well as the little guys. This may be as elaborate as action figures or as inexpensive as pens, T-shirts, or key chains. In order to get the name of your video out there, you may need to resort to trinkets (just like when you mistakenly dated Moose's wife).

5. A press junket will allow the media to use their power to get your video advertised. The cast will speak to the press and tell people how much they loved making your video. Of course, you would never tell them what to say, but you probably haven't paid them yet.

Streaming Video on the Web

Selling Yourself, Then Your Work

To paraphrase something Orson Wells once said, "We spend too much of our time trying to find funding for our work, time that could be better spent in the making of our films." Don't be concerned with "who's going to pay for this." You will gain more experience and knowledge from doing.

If reading this book only leaves you with one thought, that would be—go out and do it. Don't let anyone tell you, "You don't have the experience, knowledge, or funds." By doing it, you will get the experience and find out what works and what doesn't. Tape is inexpensive. If what you shot is junk, erase it and try again. But don't be the one that makes that decision. Your junk is someone else's treasure. Keep an open mind and get opinions from others (opinions are free; that's why everyone has one).

If you can't sell yourself and your capabilities, your videos won't have a chance of making money. I've produced a lot of garbage (it's collected every Wednesday), and I've learned along the way. Know your strengths and weakness, but only promote the strong points while secretly working on the weak.

When I sold cars for a living (I had to eat), my sales manager said, "Take what you have and sell it." If a new car in the showroom had a full-sized spare tire, I had to focus on selling that feature. "This car has a full-sized spare, not one of those useless little space-saving donuts! When a tire wears out, you have a full-sized fifth (Jack Daniels if I made the sale). When the full-sized spare is on the car, you can drive at any speed over any surface, and never worry about getting the flat fixed because you have a real tire."

If, on the other hand, the car had a space-saver spare, I had to tout that. "This car has a space-saving, weight reducing spare! Who gets a flat nowadays and wants all the weight and expense of a full-sized tire that will just rot in the truck? This tiny tire is cheaper and easier to put on and mount in an emergency on the off chance you'll ever use it." Take what you have and work with it selling yourself. "Look at the camera work in this scene. None of these actors had ever acted before I got a hold of them." Don't be egocentric and say how great you are; just point out your strengths.

Most importantly, just go ahead and shoot the video. Tools don't make the artist. The tools are now much cheaper than they have been in the past, but does that mean everything you produce will be marketable or watchable?

The Internet is one of the greatest tools for showcasing your work to the planet. Currently, the capacity for broadcasting digital images isn't up to the quality of watching the same thing on television, but everyone with computer access to the Web can view your film. At this point, this is more of a visual outlet for your video but could lead into a money-making venture. You can use the Internet to generate interest in your capabilities.

In order to sell your work, you must sell yourself to the viewer. Why should anyone want to take the time to wait for your streaming video to download? The best reason is because they read about you, and your clip seems interesting. Once you have their attention, never let it go.

The Internet and You

With everyone having access to the Internet, it was only a matter of time before budding video artists were sending their moving images over the World Wide Web. How do you get your finished video on the Web in the first place? Here's where you'll need a little help with two options: start with streaming video (or live footage) or convert the video to a streaming format. Currently, these are your only two methods available.

Streaming Live

The next section, "Is the Internet Really Ready for Me?" discusses how to take a live production and send it instantly over the Internet. But here I want to talk about the technology that enables you to convert any video into a streaming mode.

JVC has created the first Streamcorder™, the JVC GV-DV 300 that shoots "broadcast-quality video, live from your camcorder to the world." This camera was the first built to capture an image with its three *charge-coupled device* (CCD) chips and be able to output MPEG-4 streams through a wireless network onto the Internet. I received one of these cameras from JVC to review and actually sent my footage over the Internet.

If that option isn't flashy enough, the images can be recorded onto a memory card and embedded later into presentations or email messages. This is obviously the fastest way to get images in a streaming mode.

An MPEG-2 signal (broadcast-quality video) has too much information and would bog down trying to stream onto the Internet. The MPEG-4 signal has less information and enables a smoother flow. Unless you've shot or have converted to this format, you won't be on speaking terms with the Web.

I Want to Convert

The other option is to take your video and, using an editing system, change it to a web-friendly format. Jon Leland describes it best when he uses the term pseudostreaming, which is better known as a progressive download. Leland says that this conversion is "a file download that is modified to enable the viewer to see part of the clip during the process of the download. Once the viewer reaches the end of what has been transferred, he or she must wait until more content has been downloaded. Because streaming constantly caches and flushes the video stream from memory, the user never receives a copy that can be saved and copied to their hard drive."[1] This will alleviate some concern about someone copying your work and selling pirated versions. Once the video is viewed, it's gone.

Without going into a lot of detail on how streaming actually works, which is beyond the scope of this book, I will focus on what you should do to have your video look its best in this emerging medium. When shooting for streaming, keep in mind that your images will be compressed greatly so they will flow and load on computers smoothly. That means fast pans, tilts, and dollies, as well as a lot of visual information on the screen, will take more time to compress and finally open in the streaming window. When it does open, some problems may occur.

You've all seen talking heads on the Internet where the lip sync is slightly off. The video is choppy because the information has to be gathered, temporarily stored, and then displayed while more is gathering. You can wait and have the entire video store and load, but this takes eons of time.

Editing programs like Premiere, Final Cut Pro, and Avid Xpress enable you to convert your final video to a web-based movie with a single mouse click. You don't have to do anything more than prepare your HTML screen with all the pertinent information.

Without drastically changing your video, remember that the Internet will still have problems with some of your imaging. By not running at a true

---

[1]Silbergleid, Michael, and Mark. J. Pescatore. *The Guide to Digital Television, Second Edition.* Miller Freeman PSN, Inc., New York, NY 1999.

30 *frames per second* (FPS) or full-screen resolution, the image will never mirror that of your DVD version. But streaming does enable you to better sell yourself than other mediums.

It's a State of Design

Design your HTML page so when the viewer clicks on your movie icon, they can read about you while the video is loading. Having bios of the cast and crew as well as purchasing information on your DVD will fill in all of the viewer's possible blanks. Some very talented people are using the Web as a free or inexpensive method of getting their videos and themselves seen.

The trailer you created in Chapter 9, "Features," may be shown in streaming form on the Internet. This short subject will not take long to load and if viewers are excited about the trailer, you can offer your DVD for sale. This is the widest avenue of selling yourself to others. There is no greater audience than the Internet.

A group of talented computer/video people created *405: The Movie* (www.405themovie.com). This short comedy begins with a man on his way to work who gets caught behind a slow, aged driver. He ends up on a runway with a 727 landing on his truck. By taking still frames of the runway and aircraft, these geniuses mapped all the motion action in the video. Millions have watched it in the streaming mode and have purchased the video from their web site (a no-budget dream come true).

This is a perfect example of how a trailer, shown on the Internet, sold over a million copies on video because the artists sold themselves. The budget was low; they couldn't actually shoot the aircraft landing on the vehicle, but they used their knowledge of computers (and a lot of time) to create a flawless, humorous film.

What makes your videos stand out from the rest? How do you produce an amazing action picture on a tiny budget? You have an entire page to advertise why your videos are the best in the business.

Sites like iFilm (www.ifilm.com) and the Bit Screen (www.thebitscreen. com) are but two showcases for new videomakers. Videos are uploaded and made available to the masses on sites like these. Today your neighborhood, tomorrow the world (via the Internet)—how's that for a new motto?

Review: Five Ways to Sell Yourself on the Internet

1. Gain experience by going out and doing it. Never let anyone tell you that you can't do something. Many things I thought were beyond my capability proved easier once I did them. The only exception is holding your breath for three minutes—that you can't do.

2. Accentuate only your positive attributes while selling yourself to someone. Focus on strengthening your weak points so they will become positives as well. Just like working out, the more experience you gain, the better you will become. Don't just try to make your weaknesses stronger, but improve your strengths and make them better. Since I'm right-handed, I intensified my training on the left. Now I look like Popeye after eating half a can of spinach.

3. Use the Internet and get your videos streaming. This market opens up the entire planet and you may have an audience you never thought possible. Remember to use it for good rather than bad.

4. Keep fast-moving shots (pans, dollies, and tilts) to a minimum while streaming. The process is improving, but it still will take time to improve what's currently available. A short time ago we couldn't burn DVDs at an earthly price. When they plant the Internet chip in your brain, we can talk about how to access it easier.

5. Via a web page, advertise your video as well as biographies on your cast and crew. This advertising is almost as important as the streaming portion of your video. Make sure each link on your web site works. Last time I checked my bio link, I was taken to a site on glow-in-the-dark toys—I need a better last name.

Is the Internet Really Ready for Me?

Are your videos good enough for the Internet? Or is the Internet ready for you? The answer to both those questions is—yes. The first video I streamed on the Internet was to be my crowning glory. I believed it was one of my best projects, but unfortunately no one else seemed to enjoy it. I realized what was wrong and reedited the piece; now it's much more successful. It wasn't ready for the Internet.

Before sending any of your videos as streaming media on the Internet, look them over and see if they "work" in a web environment. Too many of my videos' minute details were lost in the streaming process, so much of my story was misunderstood (just like me, the story of my life). By reediting in close-ups instead of relying on the viewer seeing them within the wide shot of the frame, they were now more visible.

This is why it's important to save all your footage. The close-ups didn't work within the body of the video (on DVD) because I wanted the effect to be subtle. Always shoot cutaways so you have them in a time of need. This new streaming market made these close-ups needed. I have viewed my video hundreds of times but never thought about that happening when viewed on the Internet.

Review your videos before streaming them and see if anything is missing on the smaller screen. View your pieces at the same size and resolution that the Web allows (when you save it as video for the Web, look at it before uploading).

Multicasting

The newest word for your vocabulary is *multicasting*. The word in question means spreading the load of streaming using multiple servers in order to increase the number of available streams, thus expanding the audience. This only works for live programs or events. Jon Leland gives an example: "If one server delivers 100 streams to 100 other servers that in turn each deliver 100 streams to users, you quickly have 10,000 available streams originating from one server, but without any one server taking a major 'hit.'"[2]

On the downside, these feeds have less flexibility and are delivered or sent at a lower quality because the compression of the signal takes place in real-time instead of downloading. The technology is still there; some of the bugs just have to be worked out. Almost anyone can now be a semibroadcast facility and send live programming over the Internet.

DCTV Makes History

Housed in a turn-of-the-century fire station on the corner of Lafayette and White in Manhattan (Chinatown) is the *Downtown Community Television Center* (DCTV). Since 1972, this nonprofit media center has given the public access to electronic media arts through "artistic television." Supporting themselves financially by making documentaries for HBO, DCTV delivers live cable and Internet performances on a weekly basis.

---

[2]Ibid.

Founder Jon Alpert, believing that you need to be on the street with the people, realizes that DCTV has "very big ideas and a very low budget." Hits home, doesn't it?

Enter GlobalStreams™

"Live from Downtown," an interactive broadcast, displayed the artistic work of some extremely talented people. With GlobalStreams' GlobeCaster Studio™, the multilayered digital elements of the improvised program could be merged with live performances, virtual sets, and the audience interactivity on the Web.

With a virtual set being a must in this performance, GlobeCaster seamlessly combined the multiple backgrounds generated for the artists and performers. Operating on a 600 MHz Pentium III dual-processing system, three units were donated to DCTV in order to make their presentations and studio instruction a reality.

The "Live from Downtown" interactive broadcast was a feast for the eyes as well as a history of visual technology. Low-resolution images were input into the system and remapped onto three-dimensional shapes and reflected virtual mirrors.

The multicolored, swirling, euphoric background was placed behind the dancers as part of the virtual set. The dancers were recorded against a black or Holoset background. By using a screen material made of millions of tiny half-glass spheres, the fabric resembled a heavy, gray canvas backdrop. However, when a Holoset ring is attached to the camera's lens, the ring of blue *light-emitting diodes* (LEDs) illuminates and activates the glass spheres to bathe the fabric in a fluorescent, chroma key-like blue. The screen material may be viewed at any angle without the traditional problems of lighting a chroma key screen and you don't need flat, even lighting, and wrinkles or folds aren't an issue.

The GlobeCaster then keyed out the blue (or green) background and incorporated the virtual background of choice without the jagged or fuzzy edges that conventional chroma key may produce. The fast rendering makes the results nearly instantaneous and the talent was enclosed in a virtual environment.

Looking like a three-ring circus, all the visually charged information was presented simultaneously over the Internet. Members of the online audience (with Webcams) became part of the broadcast themselves by "wiring in" to the DCTV studio. They could watch the events transpiring live, and the studio audience could see them.

The live Internet streaming broadcast was comprised of 8 different feeds running at 128 Kb. DCTV's desire is to increase that to 300 kb in the near future. Not worried about the choppiness, some believe that the sporadic frozen frames will allow the viewer more time to digest the images.

Watch any of the news networks when they are interviewing someone in a far-off country. The interview's streaming images produced by what is known as a video phone, although delayed, allow the anchor to question correspondents all over the world without the expense or risk of sending a camera crew in volatile situations.

The Present and the Future

We've seen videos (nearly live) being broadcast from the heads of bombs as they make contact with their targets in the Gulf and Afghanistan Wars. This technology was sent to the Internet moments after it happened, just as fast as live television can make it happen.

Remember, the Internet is still evolving. With the Internet 2 in planning stages (as of this writing), this new "channel" will enable the distribution of videos and Hollywood features. The Internet as well as satellite television offers thousands of opportunities for artists to broadcast their work. Presently, not enough programming exists to fill all the space available. It looks like we need to get working!

Review: Two Ways to Make Better Streams Without Water

1. Watch your video on a small screen and see if anything is lacking when viewing it in that scale. It's best to fix anything before your program travels over the Internet. Be extremely critical and pretend you are seeing it for the first time. Does it flow (beside the choppy image)? Does it need anything else (beside clear video)? If you can read this, you are driving too close.
2. Use as many close-ups as possible to tell your story. The great director D. W. Griffith (no relation to Andy) once said, "If movie screens got any bigger, we wouldn't need close-ups." Since the Internet is a smaller screen, the opposite applies. You wouldn't want him to get mad and have Deputy Barney Fife arrest you (sorry, I still get the two confused).

Making Choppy Video Work for You

What's the most exciting thing about streaming video? Is it that your videos may be instantly seen by the masses? Is it the technology that compresses your image and then expands it again for later viewing? Or is it just the concept that you have an outlet for your creative works? Whatever the reason, the *only* drawback to this amazing distribution outlet is the choppiness. As my mother-in-law says, "This too shall pass."

Great After Effects

At the college where I teach, video students in web classes learn the operation of Adobe After Effects and make 30-second self-promotional pieces (commercials) that are uploaded to the Internet. These short bits of video don't require the use of a camera, actors, or anything else but the artist's imagination and are downright outstanding. The images can be clip art, designed logos, or bits and pieces of Photoshop or Illustrator graphics that maneuver around the screen. Generated totally in a computer environment, these videos are less jerky when viewed on the Web's displaying medium.

This is one of the best ways to keep "chop" to a minimum while streaming. The video image is the most difficult to compress, download, and play back on the computer because it's taking a medium that's not really compatible (720 × 480 is video, not computer savvy). Graphics in any program are already in the computer's format and aspect ratio, and they don't need the conversion video still desires.

This isn't necessarily a bad thing. People have been expressing themselves in video without ever holding a camera. If your masterpiece doesn't enjoy the transition to streaming, perhaps a web-generated version would fit the bill.

Popcorn Optional

When was the last time you went to see a good movie . . . on the Internet? In 1997, Nora Barry of Druid Media and Comcast (the initial title sponsor) created the Bit Screen (www.thebitscreen.com), an online movie theater that features new independent short videos from Philadelphia artists created for the Internet. What started out as just an experiment, after three months it turned into a money-making business and an outlet where digital films created for the Internet could be watched, interacted with, or simply enjoyed. Figure 10-1 is an example of the Bit Screen menu.

Having been actively involved in multimedia for the past 15 years, Nora Barry had this vision of an Internet site for quite some time. She came up with the original concept and design, and she recently had Arts Tribe, Inc. redesign a better, more functional web site.

Because bit-streamed "movies" over the Internet had been quite new at the time, the Bit Screen was the only outlet available to Internet video-makers. The bit streaming of the images enables the Internet user to begin watching the movie as it is loaded, rather than waiting for the entire file to download.

Figure 10-1
The Bit Screen

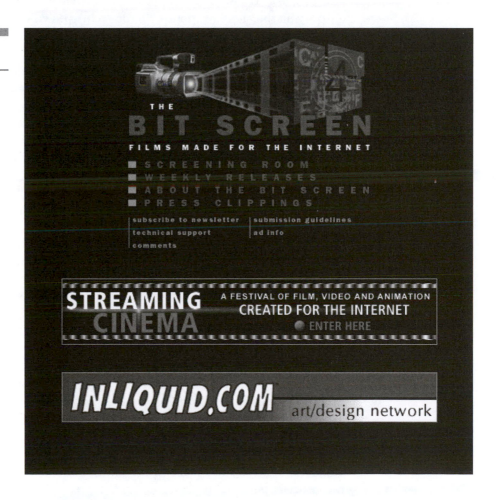

History in the Making

What started out as an outlet for Philadelphia-area artists has bloomed into an international web site receiving more than 400,000 hits each week by the late 1990s. These films were never intended for viewing on the movie screen or on video. Barry wanted to take these short pieces from the movie theater to a real channel, all while still on the Internet. Since all the films are conceived and created to arrive by modem, most of these artistic works have no outlet other than the computer screen. Once on the computer screen, the entire world has access without the problems of international standards conversions.

Even though the current technology is inconsistent and bit streaming is still evolving, the web site patron can still view the short films in their entirety on the small (computer) screen. Once downloaded, the images are slightly choppy. Video, normally running at 30 FPS, keeps movement fluid, and images on the Bit Screen run at 16 FPS. In film school, we were told that "persistence of vision" is what allows a viewer to perceive motion from a series of still frames strung together. This means that the greater the number of frames, the smoother the perceived motion. Although clear and sharp, these streamed moving images are not up to television's broadcast standards. But the main thrust is to get the artist's work seen.

This has been the goal of the site. An artist must have his or her work seen in order to receive feedback. This feedback is crucial to one's development as an artist and videomaker. That's why the quality (one chip or three, VHS or digital, simple storytelling or effects-laden extravaganzas) isn't important. The viewer in this outlet is not focusing on the equipment used, but rather the talent behind it. The Internet is really the only outlet where people can center just on this aspect. This is where the great videos stand out from the rest.

The process also fits nicely with no-budget production. It costs virtually nothing to upload a video onto the server (just your time and any Internet connection costs). If your project fails because "it needs something," it can be easily changed unlike a completed video that has already been burned to DVD. Everyone wants a great piece to be even better with the input of others. Just like the Army, you want your video to be all it can be. With the mindset of "what kind of camera did you use?" not being an issue, the video artist can put all of his or her creativity into the program and getting it seen.

The quality will improve as the technology does. Because G2 technology may not be accessible to every viewer, the Bit Screen has free downloadable options for viewing all the films. Currently, filmmakers submitting work to Barry aren't paid, but their work *is* getting seen. This is another great way to get ideas for future projects. By watching the work of others, a spark may be ignited. Don't copy their work (imitation is not the most sincere form of flattery in this case), but build upon it.

Advertisers support funding for the Bit Screen. "As the Internet grows, so will the web site," says Barry. What started out as bimonthly and became a monthly site, Barry now features new films every week. This concept keeps the site constantly fresh while continuing to offer an outlet for budding digital filmmakers.

Creating short videos on the computer allows the artist to remake the videos after they are finished, and even still remake them while they are being shown. Only the Internet offers this type of interactivity to the viewer. "Employing technology this way turns the Internet into a storyteller," says Barry. Although currently at the cutting edge of technology, these two- to seven-minute epics only pause long enough on your computer screen to allow additional information to be collected.

Other filmmakers (a term often used to describe people who shoot video) have also pushed this new technology to the limits. Visual and audio tracks are shuffled and reassembled each time they are viewed to tell a different story. Try to do that in a conventional video!

Barry is also attracting a lot of interest from Europe. Most of the short pieces that Barry receives for possible inclusion in the Bit Stream are digital based, submitted on mini-DV tape, Windows (Real) Media, or as a Quick-Time file.

When asked where she sees the Bit Screen will be going in the future, Barry told me she believes it "will evolve into a computer channel, not unlike a channel on television, basically what the Internet 2 will be." It could also be called an entertainment Internet channel with video on demand. "The content of the programming on my channel will change as technology changes." With television being around since the 1940s, the Internet is really still in its infancy. Just like early programming, this new marketplace has an incredible need for new material. With thousands of channels available on TV through satellite, this lack of programming is a real need.

One of the programs recently completed is with the Big Picture Alliance. This nonprofit organization allows inner-city kids to make Internet films about their lives. These young animators and video artists are channeling their creative energy into expressing themselves in short bitstream movies. The Internet has now become the storyteller. One person can start the story and someone else can add to it with images of their own. In much the same way that stories were told for hundreds of years, through digital technology they can be kept fresh in the retelling.

Whether you want to interact with your movie or prefer to watch passively, the Bit Screen films can be used as well as watched. It's there as much for the spectator as it is for the film's creator.

Just the Beginning

Technology will change quickly whether we want it to or not. Soon everything will be shot on a digital, high-definition medium that probably will not be magnetic tape based. The future may have us shooting in three-dimensional imaging or perhaps in holograms.

Whatever the format of camera (analog, digital, high-definition, one-chip, or three-chips), the type of recording media (tape, disc, chip, card, or hologram), the editing process (linear, nonlinear, or imagined), or the distribution process (tape, disc, or Internet), the most important aspect is the story and how you make that a reality with spending no money—that will never change with the passage of time.

APPENDIX

Magazines

American Cinematographer
1782 North Orange Drive
Hollywood, CA 90028
(800) 448-0145
www.cinematographer.com

AV Multimedia Producer
701 Westchester Avenue
White Plains, NY 10604-3098
(800) 800-5474
www.avvmmp.com

Broadcast Engineering
P.O. Box 12914
Overland Park, KS 66282-2914
(913) 967-1903 (Fax)
www.broadcastengineering.com

Digital Cinema
460 Park Avenue South
9th Floor
New York, NY 10016
(212) 378-0400
www.digitalcinema.com

Digital Television
460 Park Avenue South
9th Floor
New York, NY 10016
(212) 378-0400
www.digitaltelevision.com

Digital Video
600 Harrison Street
San Francisco, CA 94107
(415) 947-6000
www.dv.com

E-Media
213 Danbury Road
Wilton, CT 06897-4007
(800) 248-8466
www.emedialive.com

Film & Video
701 Westchester Avenue
White Plains, NY 10604
(914) 328-9157
www.filmandvideomagazine.com

Government Video
460 Park Avenue South
9th Floor
New York, NY 10016
(212) 378-0400
www.governmentvideo.com

Millimeter
5 Penn Plaza
13th Floor
New York, NY 10001
(913) 341-1300
www.millimeter.com

Mix
6400 Hollis Street
Suite 12
Emeryville, CA 94608
(510) 653-3307
www.mixonline.com

Post
25 Willowdale Avenue
Port Washington, NY 11050
(516) 767-2500
www.post.com

TV Technology
5827 Columbia Pike
Suite 310
Falls Church, VA 22041
(703) 998-7600
www.tvtechnology.com

Videography
460 Park Avenue South
9th Floor
New York, NY 10016
(212) 378-0400
www.videography.com

GLOSSARY

Following are definitions of some of the "new" words I used throughout the book. Each description is mine and written in my own words. The Webster's definitions would be more precise, but mine are hopefully simple enough that I could understand them. This is by no means an all inclusive listing, but it does give definitions of words that are used in digital filmmaking.

Authoring When creating a DVD menu, authoring is the process of writing and encoding a "script" for the branching path. When the user activates a menu, the branching takes them to the correct destination. Writing and creating this process is called authoring.

Axis An imaginary line drawn between the cameras and the "action" that the videographer cannot cross.

Beam splitter a partially silvered mirror that allows some of the light to past through it while reflecting 45 percent back to the camera.

Blood squibs sacks of stage blood containing a small explosive charge. When detonated, the charge ruptures the sack and blood is released.

Branching the path the user takes moving from one screen to another.

CCD charged coupled device that collected the light entering a video camera.

CCU central control unit or the "brains" of a computer or camera.

Chimera a white fabric enclosed light that contains a powerful light source that diffuses or softens.

CTB color temperature blue. A gel that adds blue colored light.

CTO color temperature orange. A gel that adds orange colored light.

Cyc a seamless background or backdrop that curves from the walls to the floor giving the illusion or a smooth, continuous background

Daylight the light the sun produces at a color temperature of 5600 degrees Kelvin.

Dichromatic filters a glass filter placed in front of a light that changes its color temperature from tungsten to daylight.

Digitize transferring video footage from your camera or deck into your computer's storage device.

DLT digital linear tape that holds information to be used in creating a glass master for DVD replication.

Doubler a lever that adds a two times (2X) magnifier to your lens, doubling its focal length. A 50mm lens with a doubler becomes a 100mm lens.

Encoding   transferring video footage with time code in and out points from your camera or deck into a DVD authoring system.

Foot-candle   the amount of light one candle gives off at a distance of one foot. Sixty foot-candles would be the amount of light 60 candles gives at a distance of one foot.

Gigabyte   1,000,000 bytes of information or 1000 megabytes used in storage.

Hard sell   a type of commercial that crams information down a viewer's throat. Usually done though loud voice-over and repeating the product's name too many times.

Hostess tray   a car mount that attaches to the doors of an automobile much like a hostess would use when serving food in a drive-in.

Jimmy Jib   a boom mounted camera that allows vertical, horizontal, and lateral movement.

Kelvin   a color temperature scale measured in degrees (Kelvin) with tungsten being 3200° and daylight being 5600°.

Lipstick camera   a tiny camera roughly the size of a tube of lipstick.

MPEG-2   motion pictures expert group. The format of video and compression suitable for digital broadcast television (NTSC, 525 lines, 30fps).

MPEG-4   motion pictures expert group. The format of video and compression suitable for streaming video (320 × 240, 15fps).

Multicasting   spreading the load of streaming images using multiple servers in order to increase the number of available streams . . . all at the same time.

NTSC   National Television Standards Committee, the broadcast standard in North American and Japan since the 1940s (525 lines of horizontal resolution, 60 Hz).

PAL   Phase Alteration of Line. The broadcast standard in Europe (625 lines of horizontal resolution, 50 Hz).

Paper cut   basic video editing using information written on paper. The in and out points of each are written, and the information on the paper is cut and pasted together to perform a rough edit.

Persistence of vision   a blurring of images that the eye and brain create to simulate motion. Any image moving faster than nine frames per second is perceived as a moving image because the first image blurs slightly into the next.

Polarizing filter   a filter, when rotated, that removes some of the ultra-violet light and glare from reflective surfaces.

POV   point of view shot. A camera angle meant to duplicate what a person or object sees from their perspective.

Prospectus   a detailed form that lists budgeting information to present to a perspective investor in the hopes of gaining funds.

Pseudo streaming   a progressive downloading of screen information. As information is gathered, it is display on the screen instantly.

Pushpin principle   using thumbtacks to hold sheets of diffusion material to the studs on a 2 × 4 set.

Release   a form that is signed by on-camera talent allowing them to use their likeness in your productions.

Soft sell   a type of commercial that delivers the product information is delivered in a gentle, calm, and appealing way. The opposite of hard sell.

Streaming   video, audio, text or animation that is displayed on the Internet, broadcast network, or local storage.

Swatch booklet   a booklet containing samples of color gels and diffusion material.

Time code   A number displayed in hours, minutes, seconds, and frames given to every frame of video to assist in editing.

XLR   a connector, also know as a Canon connector (because Canon Corporation developed it) that professional audio people use to transfer sound via a three pin and wire cable. Professionals use an XLR cable to carry sound; consumers use RCA cables to do the same thing.

INDEX

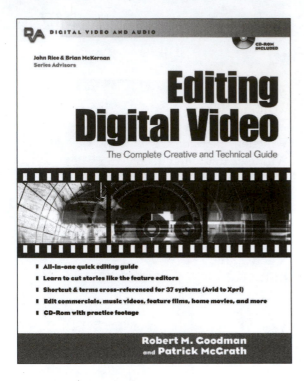

Editing Digital Video

Robert Goodman and Patrick McGrath
With forward by Bob Turner
361 pages/ $39.95 /Softcover /0–07-1406352

"—it provides sound technical advice and instruction as well as excellent guidance in aesthetic and formal issues. I'd recommend it to anyone seeking to get into video editing and even to more advanced editors looking for new ideas."
Videographer and Professor Christopher Pavsek on Amazon.com

With *Editing Digital Video*, you'll finally unleash your creativity. Learn more in one session than you would from any user manual. Experts Goodman and McGrath share their insights and explain the tools—from iMovie or Premiere to appliances like Casablanca and Screenplay or professional systems such as Avid, Discreet, and Media 100.

THE DIGITAL WAY TO CUT VIDEO—Quickly acquire the skills you need to:
• Edit features, commercials, documentaries, and music videos.
• Cut your stories like features editors
• Use Pro Tips for home movies
• Create Titles and Effects
• Work with video, DVD, and Web-based media
• Take advantage of proven techniques from the pros

PRACTICE WITH ACTUAL FOOTAGE ON A CD-ROM TOOLKIT.

For the **amateur**, turn your family videos into stories. For the **professional**, learn to cut your films using the latest digital video tips and tricks.